World Statistics on Mining and Utilities

2016

Statistics Division
United Nations Industrial Development Organization
P.O. Box 300
A-1400 Vienna, Austria
Fax: +43-1-26026-6802
Email: stat@unido.org
www.unido.org/statistics

World Statistics on Mining and Utilities 2016

UNITED NATIONS INDUSTRIAL DEVELOPMENT ORGANIZATION
VIENNA
2016

Published by
Edward Elgar Publishing Limited
The Lypiatts
15 Lansdown Road
Cheltenham
Glos GL50 2JA
UK

Edward Elgar Publishing, Inc.
The William Pratt House
9 Dewey Court
Northampton
Massachusetts 01060-3815
USA

A catalogue record for this book
is available from the British Library

This book is available electronically in the **Elgar**online
Economics subject collection
DOI 10.4337/9781786433473

The description and classification of countries, economies and territories in this publication and the arrangement of the material do not imply the expression of any opinion whatsoever on the part of the Secretariat of the United Nations Industrial Development Organization concerning the legal status of any country, territory, city or area, or of its authorities, or concerning the delimitation of its frontiers or boundaries, or regarding its economic system or degree of development.

The designation "country, economy or area" covers countries, economies, territories, cities and areas. The designation "region, country, economies or area" extends the preceding one by including groupings of entities designated as "country, economy or area". Designations such as "industrialized" and "developing and emerging industrial" economies are intended for statistical convenience and do not necessarily express a judgement about the stage reached by a particular country, economy or area in the development process.

FSC www.fsc.org MIX Paper from responsible sources FSC® C013604

ISBN 978 1 78643 346 6 (cased)
ISBN 978 1 78643 347 3 (ebook)

Printed and bound by CPI Group (UK) Ltd, Croydon, CR0 4YY

ABOUT THIS PUBLICATION

This is the fourth issue of the World Statistics on Mining and Utilities. Following its first release in 2010 the publication was distributed to the National Statistical Offices (NSOs), international organizations and other data users. UNIDO also informed the forty-second session of the United Nations Statistical Commission about the publication with a brief note on its sources and methods and periodicity. UNIDO sincerely appreciates the cooperation with NSOs, OECD (Organisation for Economic Co-operation and Development), UNSD (United Nations Statistics Division) and other international organizations in providing data for this publication.

Wide interest expressed by international data users, especially knowledge institutions and development partners, in mining and utilities statistics confirms the relevance of this publication. Particular interest was shown in statistics on energy related sectors such as production of coal, crude petroleum, natural gas as well as the production and distribution of electricity and gas.

The publication was prepared by a team of statistical staff under the overall supervision of Shyam Upadhyaya, Chief of Statistics Division of UNIDO. Valentin Todorov developed the software system and produced the final manuscript. Kateryna Gumeniuk, P.Y. Rita Liang, Regina McFerren-Zenz and Romana Ransmayr completed the data preparation and the compilation of the country tables. Anja Sedic collected value added data from international sources and prepared the statistical tables for part I.

The team would also like to express its sincere thanks to Ludovico Alcorta, Director of the Department of Policy Research and Statistics of UNIDO, for his support in bringing out this publication.

UNIDO has granted the exclusive publishing rights of the *World Statistics on Mining and Utilities* to Edward Elgar Publishing Limited. The continued partnership with Edward Elgar Publishing Limited on UNIDO statistical publications is greatly appreciated.

List of abbreviations and acronyms:

CIS	Commonwealth of Independent States
ECOWAS	Economic Community of West African States
EU	European Union
GCC	Gulf Cooperation Council (consisting of Bahrain, Kuwait, Oman, Qatar, Saudi Arabia, United Arab Emirates)
GDP	gross domestic product
ISIC	International Standard Industrial Classification of All Economic Activities
MVA	manufacturing value added
n.e.c.	not elsewhere classified
NSOs	National Statistical Offices
OECD	Organisation for Economic Co-operation and Development
OPEC	Organization of the Petroleum Exporting Countries
UNIDO	United Nations Industrial Development Organization
UNSD	United Nations Statistics Division

CONTENTS

Part II. COUNTRY TABLES

DATA AVAILABLE ONLINE AND ON CD-ROM

Database	Description
INDSTAT2: UNIDO Industrial Statistics Database at the 2-digit level of ISIC code (Revision 3)	Long time series data for more than 160 countries/areas compiled on eight principle indicators (number of establishments, total employment, female employment, wages and salaries paid to employees, output, value added, gross fixed capital formation and index numbers of industrial production) starting from 1963 onwards. The data are arranged at the 2-digit level of ISIC (Revision 3) pertaining to the manufacturing sector, which comprises 23 manufacturing divisions.
INDSTAT4: UNIDO Industrial Statistics Database at the 3- and 4-digit levels of ISIC code (Revision 3 & 4)	Time series data for more than 130 (Revision 3)/70 (Revision 4) countries/areas on seven selected data items including number of establishments, total employment, female employment, wages and salaries paid to employees, output, value added and gross fixed capital formation. The data are arranged at the 3- and 4-digit levels of ISIC (Revision 3 & 4) pertaining to the manufacturing sector.
MINSTAT: UNIDO Mining & Utilities Statistics Database at the 2- and 3-digit levels of ISIC code (Revision 3 & 4)	Time series data for more than 100 (Revision 3)/50 (Revision 4) countries/areas on seven selected data items including number of establishments, total employment, female employment, wages and salaries paid to employees, output, value added and gross fixed capital formation. The data are arranged at the 2- and 3-digit levels of ISIC (Revision 3 & 4) pertaining to the mining and utilities sectors.
IDSB: UNIDO Industrial Demand-Supply Balance Database at the 4-digit level of ISIC code (Revision 3 & 4)	Time series data for more than 120 (Revision 3)/60 (Revision 4) countries/ areas on output, imports, exports and apparent consumption (all in current United States dollars) at the 4-digit level of ISIC Revision 3 & 4.
UNIDO Manufacturing Value Added (MVA) Database	Country data on GDP, MVA and population for various regions and groups of economies displayed at constant 2010 and at current United States dollars prices.
UNIDO CIP Database	Country data on ranks and values of Competitive Industrial Performance (CIP) index and its component indicators.

Further details on above products may be obtained from:
Statistics Division
UNIDO
Email: stat@unido.org
Internet: http://www.unido.org/statistics

1 INTRODUCTION

The Statistics on Mining and Utilities are compiled using data collected by UNIDO from the National Statistical Offices as well as those obtained from the Organisation for Economic Co-operation and Development for OECD member countries. The publication intends to present the detail country data on major indicators of statistics on mining and utilities at 2-digit and/or 3-digit level of International Standard Industrial Classification of All Economic Activities (ISIC), Revision 3 and Revision 4.

The main purpose of this publication is to provide internationally comparable statistics on mining and utilities sectors on major indicators. Concepts and methods applied to this publication, which are fully compatible to those used in the *International Yearbook of Industrial Statistics*, are briefly described below.

1.1 Scope and coverage

According to the International Recommendations for Industrial Statistics 2008, industrial statistics are those related to the economic activities defined under following sections of ISIC Revision 4:

Section B: Mining and quarrying
Section C: Manufacturing
Section D: Electricity, gas, steam and air conditioning supply
Section E: Water supply; sewerage, waste management and remediation activities

Data pertaining to Section C are published in the *International Yearbook of Industrial Statistics* (latest edition 2016). This publication complements the latter and refers to Sections B, D and E, which are broadly understood as mining and utilities sectors. In ISIC Revision 3, these sectors correspond to following categories:

Category C: Mining and quarrying
Category E: Electricity, gas and water supply

Details of the industry groups under these sections and categories with 2-digit and 3-digit codes are given in appendix II.

The primary source of data at the country level is mainly the annual industrial survey which defines the coverage of data for the reporting country.

1.2 Sources and methods

Part I of the publication represents combined value added data for mining and utilities sectors. These data were obtained from various national and international sources including the United Nations Statistics Division (UNSD). Data from the external sources were supplemented with estimates of UNIDO statisticians.

The country tables presented in part II are compiled from the data directly collected from NSOs except for the OECD member countries, for which data were collected by OECD and provided to UNIDO. Information is solicited through a questionnaire issued jointly by the two organizations.

1.3 Part I. Summary tables

Value added of mining and utilities are estimated in accordance with the national accounting concept. In this part, the sum of total value added of Sections B, D and E (Revision 4) and Categories C and E (Revision 3) are presented for different country groups and years. These values indicate the actual contribution of mining and utilities sectors to GDP of a country.

Country groupings, listed in appendix I, are arranged according to the established classification practices in other UNIDO statistical publications.

1.4 Part II. Country tables

Part II presents tables for the major indicators of statistics on mining and utilities, namely: number of establishments, number of employees (or, if not reported, number of persons engaged), wages and salaries paid to employees, output and value added. All value data are presented in current national currencies.

1.4.1 Statistical unit

For most countries and areas, data presented in this publication represent the activity of "establishments" classified to specific industry as recommended in the *International Standard Industrial Classification of All Economic Activities,* Revision 3 and Revision 4. In few cases, statistical unit is "enterprise", "local unit" or "kind of activity unit". An "establishment" is ideally defined as an economic unit that engages, under a single ownership or control, in one, or predominantly one, kind of activity at a single physical location. An "enterprise" is a legal entity possessing the right to conduct business in its own name; for example, to enter into contracts, own property, incur liability for debts, and establish bank accounts. An enterprise may comprise one or more establishments with different locations.

1.4.2 Reference period

The statistics in the tables relate, in general, to the calendar year. It should be noted, however, that in many cases where the basic reference period of the industrial inquiry is the calendar year, returns covering proximate fiscal years may be accepted for reporting purposes and the data for these years incorporated in the calendar-year aggregate without adjustment. In a few countries, fiscal years normally used for public accounting purposes have been adopted as the basic reference periods. In the case of fiscal-year coverage, the year indicated in the tables refers to the calendar year in which the major part of the fiscal year falls. In the case of fiscal years from 1 July to 30 June, the year referred to is normally the one in which the fiscal year ends.

1.4.3 Description of data items

Data items are described in accordance with the International Recommendations for Industrial Statistics. All values are in national currency units and are at current prices unless otherwise indicated.

(1) Number of persons engaged and number of employees

The number of persons engaged is defined as the total number of persons who worked in or for the establishment during the reference year. The concept covers working proprietors, active business partners and unpaid family workers as well as employees, but excludes home workers. The figures reported refer normally to the average number of persons engaged during the reference year, obtained as the sum of the "average number of employees" during the year and the total number of other persons engaged measured for a single period of the year. The category "employees" is intended to include all persons engaged other than working proprietors, active business partners and unpaid family workers. In this publication, preference has been given, whenever possible, to employees over persons engaged.

(2) Wages and salaries

Estimates of wages and salaries include all payments in cash or in kind made to "employees" during the reference year in relation to work done for the establishment. Payments include: (a) direct wages and salaries; (b) remuneration for time not worked; (c) bonuses and gratuities; (d) housing allowances and family allowances paid directly by the employer; and (e) payments in kind.

Compensation of employees is equivalent to wages and salaries plus employers' contributions on behalf of their employees paid to social security, pension and insurance schemes, as well as the benefits received by employees under these schemes and severance and termination pay.

(3) Output

The measure of output normally used in the tables is the census concept that covers only activities of an industrial nature. The value of census output in the case of estimates compiled on a production basis comprises: (a) the value of sale of all products of the establishment; (b) the net change between the beginning and the end of the reference period in the value of work in progress and stocks of goods to be shipped in the same condition as received; (c) the value of industrial work done or industrial services rendered to others; (d) the value of goods shipped in the same condition as received less the amount paid for these goods; and (e) the value of fixed assets produced during the period by the unit for its own use. In the case of estimates compiled on a shipment basis, the net change in the value of stocks of finished goods between the beginning and the end of the reference period is also included.

Gross output is equivalent to census output plus the revenue from activities of a non-industrial nature. Valuation methods differ from country to country. An increasing number of countries are reporting data at basic prices, which excludes taxes on commodity and includes commodity related subsidies. Other methods of reporting valuation are: (a) at factor costs, which excludes all indirect taxes falling on production and includes all current subsidies received in support of production activities; and (b) at producers' prices, which includes all indirect taxes except value added tax or any other deductible taxes and excludes all subsidies.

(4) Value added

The measure of value added normally used in the tables is the census concept, which is defined as the value of census output less the value of census input. Items covered in the latter include: (a) value of materials and supplies for production (including cost of all fuels and electricity purchased); and (b) cost of industrial services received (mainly payments for contract and commission work and repair and maintenance work). If input estimates are compiled on a “received” rather than on a “consumed” basis, the result is adjusted for the net change between the beginning and the end of the period in the value of stocks of materials, fuel and other supplies.

Total value added is the national accounting concept. It is ideally represented by the contribution of the establishments in each branch of activity to the gross domestic product. For the measure of total value added, the cost of non-industrial services is deducted and the receipts for non-industrial services are added to census value added. The estimates, whether in terms of census value added or total value added, are gross of depreciation and other provisions for capital consumption, unless otherwise stated. The valuation may be at factor costs, at basic prices or at producers’ prices, depending on the treatment of indirect taxes and subsidies as described above.

Appendix I

LIST OF COUNTRIES AND AREAS INCLUDED IN SELECTED GROUPINGS

Industrialized Economies

EU [a/]

Austria
Belgium
Czech Republic
Denmark
Estonia
Finland
France
Germany
Hungary
Ireland
Italy
Lithuania
Luxembourg
Malta
Netherlands
Portugal
Slovakia
Slovenia
Spain
Sweden
United Kingdom

Other Europe

Andorra
Iceland
Liechtenstein
Monaco
Norway
Russian Federation
San Marino
Switzerland

East Asia

China, Hong Kong SAR
China, Macao SAR
China, Taiwan Province
Japan
Malaysia
Republic of Korea
Singapore

West Asia

Bahrain
Kuwait
Qatar
United Arab Emirates

North America

Bermuda
Canada
Greenland
United States of America

Others

Aruba
Australia
British Virgin Islands
Cayman Islands
Curaçao
French Guiana
French Polynesia
Guam
Israel
New Caledonia
New Zealand
Puerto Rico
United States Virgin Islands

a/ Excluding non-industrialized EU economies.

(continued)

Appendix I

LIST OF COUNTRIES AND AREAS INCLUDED IN SELECTED GROUPINGS

Developing and Emerging Industrial Economies by Region

AFRICA

Central Africa

Cameroon
Central African Republic
Chad
Congo
Equatorial Guinea
Gabon
Sao Tome and Principe

Eastern Africa

Burundi
Comoros
Djibouti
Eritrea
Ethiopia
Kenya
Réunion
Rwanda
Somalia
Uganda

North Africa

Algeria
Egypt
Libya
Morocco
South Sudan
Sudan
Tunisia

Southern Africa

Angola
Botswana
Democratic Rep of the Congo
Lesotho
Madagascar
Malawi
Mauritius
Mozambique
Namibia
Seychelles
South Africa
Swaziland
United Republic of Tanzania
Zambia
Zimbabwe

Western Africa

Benin
Burkina Faso
Cabo Verde
Côte d'Ivoire
Gambia
Ghana
Guinea
Guinea-Bissau
Liberia
Mali
Mauritania
Niger
Nigeria
Senegal
Sierra Leone
Togo

ASIA & PACIFIC

China

Central Asia

Kazakhstan
Kyrgyzstan
Mongolia
Tajikistan
Turkmenistan
Uzbekistan

South Asia

Afghanistan
Bangladesh
Bhutan
India
Maldives
Nepal
Pakistan
Sri Lanka

South East Asia

Brunei Darussalam
Cambodia
Indonesia
Lao People's Dem Rep
Myanmar
Philippines
Thailand
Viet Nam

West Asia

Armenia
Azerbaijan
Iran (Islamic Republic of)
Iraq
Jordan
Lebanon
Oman
Saudi Arabia
State of Palestine
Syrian Arab Republic
Yemen

Other Asia & Pacific

Cook Islands
Democratic People's Rep of Korea
Fiji
Kiribati
Marshall Islands
Micronesia, Federated States of
Palau
Papua New Guinea
Samoa
Solomon Islands
Timor-Leste
Tonga
Tuvalu
Vanuatu

EUROPE

Albania
Belarus
Bosnia and Herzegovina
Bulgaria
Croatia
Cyprus
Georgia
Greece
Latvia
Montenegro
Poland
Republic of Moldova
Romania
Serbia
The f. Yugosl. Rep of Macedonia
Turkey
Ukraine

LATIN AMERICA

Caribbean

Anguilla
Antigua and Barbuda
Bahamas
Barbados
Cuba
Dominica
Dominican Republic
Grenada
Guadeloupe
Haiti
Jamaica
Martinique
Montserrat
Saint Kitts and Nevis
Saint Lucia
Saint Vincent and the Grenadines
Trinidad and Tobago

Central America

Belize
Costa Rica
El Salvador
Guatemala
Honduras
Mexico
Nicaragua
Panama

South America

Argentina
Bolivia (Plurinational State of)
Brazil
Chile
Colombia
Ecuador
Guyana
Paraguay
Peru
Suriname
Uruguay
Venezuela (Bolivarian Republic of)

Appendix II

Detailed description of International Standard Industrial Classification of All Economic Activities (ISIC) - Revision 3

ISIC	Description
C	**MINING AND QUARRYING**
Division 10	**Mining of coal and lignite; extraction of peat**
101	Mining and agglomeration of hard coal
102	Mining and agglomeration of lignite
103	Extraction and agglomeration of peat
Division 11	**Extraction of crude petroleum and natural gas; service activities incidental to oil and gas extraction, excluding surveying**
111	Extraction of crude petroleum and natural gas
112	Service activities incidental to oil and gas extraction excluding surveying
Division 12	**Mining of uranium and thorium ores**
120	Mining of uranium and thorium ores
Division 13	**Mining of metal ores**
131	Mining of iron ores
132	Mining of non-ferrous metal ores, except uranium and thorium ores
Division 14	**Other mining and quarrying**
141	Quarrying of stone, sand and clay
142	Mining and quarrying n.e.c.
E	**ELECTRICITY, GAS AND WATER SUPPLY**
Division 40	**Electricity, gas, steam and hot water supply**
401	Production, collection and distribution of electricity
402	Manufacture of gas; distribution of gaseous fuels through mains
403	Steam and hot water supply
Division 41	**Collection, purification and distribution of water**
410	Collection, purification and distribution of water

Appendix II (continued)

Detailed description of International Standard Industrial Classification of All Economic Activities (ISIC) - Revision 4

ISIC	Description
B	**MINING AND QUARRYING**
Division 05	**Mining of coal and lignite**
051	Mining of hard coal
052	Mining of lignite
Division 06	**Extraction of crude petroleum and natural gas**
061	Extraction of crude petroleum
062	Extraction of natural gas
Division 07	**Mining of metal ores**
071	Mining of iron ores
072	Mining of non-ferrous metal ores
Division 08	**Other mining and quarrying**
081	Quarrying of stone, sand and clay
089	Mining and quarrying n.e.c.
Division 09	**Mining support service activities**
091	Support activities for petroleum, natural gas extraction
099	Support activities for other mining and quarrying
D	**ELECTRICITY, GAS, STEAM AND AIR CONDITIONING SUPPLY**
Division 35	**Electricity, gas, steam and air conditioning supply**
351	Electric power generation, transmission and distribution
352	Manufacture of gas; distribution of gaseous fuels through mains
353	Steam and air conditioning supply
E	**WATER SUPPLY; SEWERAGE, WASTE MANAGEMENT AND REMEDIATION ACTIVITIES**
Division 36	**Water collection, treatment and supply**
360	Water collection, treatment and supply
Division 37	**Sewerage**
370	Sewerage
Division 38	**Waste collection, treatment and disposal activities; materials recovery**
381	Waste collection
382	Waste treatment and disposal
383	Materials recovery
Division 39	**Remediation activities and other waste management services**
390	Remediation activities and other waste management services

Part I
SUMMARY TABLES

An overview of the world's mining and utilities

The mining and utilities sectors comprise mining quarrying activities and production, as well as the supply of electricity, gas and water, waste collection, disposal and remediation activities, broadly referred to as utilities. Mining and quarrying activities produce essential raw materials for further processing by the manufacturing industry, while the utilities sector's products and services are directly supplied to consumers via a permanent infrastructure of lines, mains and pipes. The scope of the utilities sector has expanded in Revision 4 of the International Standard Industrial Classification of All Economic Activities to include the supply of hot water, steam and air-conditioning.

The production activities of mining and quarrying include the extraction of minerals that occur naturally such as solids (coal and ores), liquids (petroleum) or gases (natural gas), as well as quarrying of stone, sand, clay and various other materials. The current edition of World Statistics on Mining and Utilities highlights the growth and structure of these sectors at country, regional and global level. The mining and utilities sectors' figures for value added and GDP are rebased from 2005 to 2010 following the International Yearbook of Industrial Statistics 2016. This change has created a new data series at a more recent price level.

The overall trend observed in the mining and utilities sectors is influenced by an increasing amount of consumption of utility products and services. However, the share of mining activities remains dominant in many countries. The growth of value added of mining activities is generally low, which is attributed to mainly two reasons. First, the gradual depletion of mineral resources has reduced the growth potential of mineral products in many countries. Secondly, falling crude oil prices have significantly influenced its production value, which accounts for significant part of mining and quarrying value added in several countries. The output of mining activities in recent years has grown at a modest rate, even in GCC countries which have been major suppliers of petroleum products in the world market for many years. The combined value added of mining and utilities grew at an average rate of 2.76 per cent per annum in Saudi Arabia and 2.98 per cent in the United Arab Emirates in 2003-2015. GCC countries have also made some progress in terms of economic diversification, thereby reducing their economies' dependence on the mining sector. Similarly, the value added of the mining and utilities sectors droped by 3.8 per cent in CIS countries, which is another major oil producing region in the world.

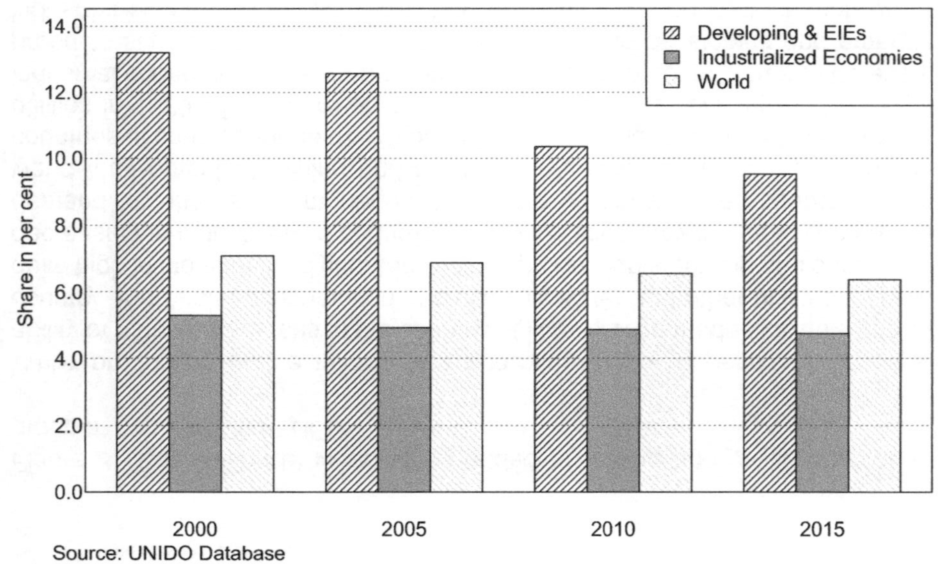

Figure 1: Share of value added of the mining and utilities sectors in GDP at constant 2010 USD

The combined contribution of the mining and utilities sectors to the total economy in 2015 was estimated at 6 per cent at the world level. The share of these activities varies significantly by country due to the sparse distribution of minerals across the globe. Mining and quarrying play a leading role in the natural resource-based economies of the developing world. Their share in GDP is close to 50 per cent in Angola, Equatorial Guinea and Congo. The mining sector of some economies of the Gulf region accounts for the substantial part of the economy, despite a decrease in recent years. In general, the share of mining and utilities, as illustrated in Figure 1, is declining in developing as well as in industrialized economies.

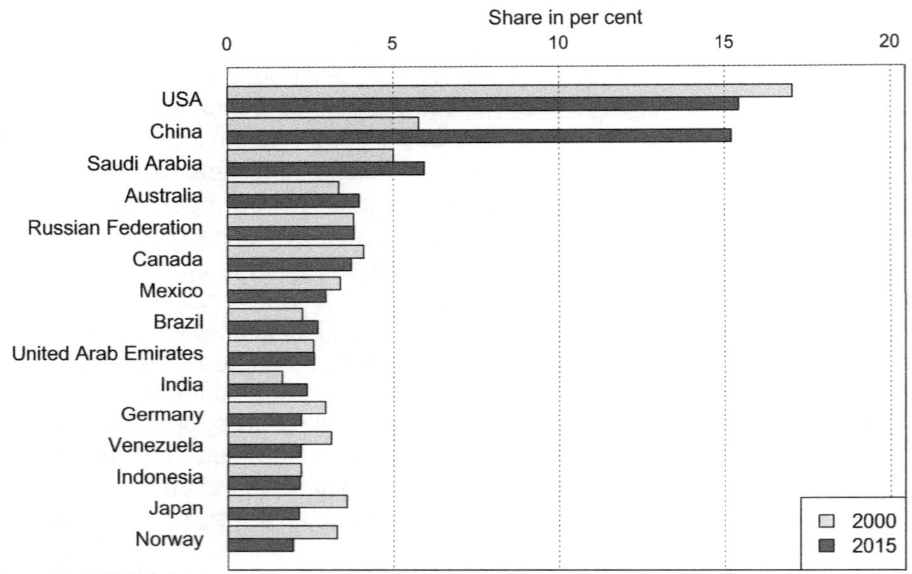

Source: UNIDO Database

Figure 2: Share of major economies in world total value added of mining and utilities at constant 2010 USD

The U.S. and China account for more than one-fourth of the mining and utilities sectors' world total value added. The share of the mining and utilities sectors in world production has significantly decreased since 2000 in major industrialized countries such as Canada, Germany, Japan and the U.S. (Figure 2). By contrast, the share of these sectors in world production has increased for countries such as Brazil, China and the United Arab Emirates. Although China currently ranks second after the U.S., its position relative to population size is much lower. In 2015, Qatar, Kuwait, Norway, Brunei Darussalam and Equatorial Guinea topped the ranks in terms of the per capita value added of the mining and utilities sectors. The per capita value added of mining and utilities is higher in industrialized countries on account of the utilities sector. Although production in the mining and quarrying sector is declining in these countries, the supply of electricity, gas and water is rising.

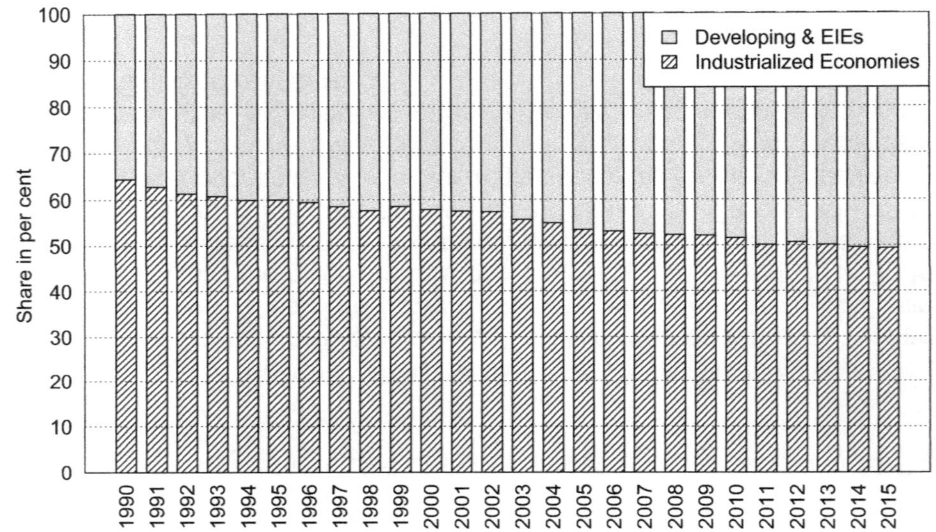

Source: UNIDO Database

Figure 3: Distribution of world value added of mining and utilities by country group at constant 2010 USD

While production in the mining and quarrying sector depends on the availability of natural resources in the country, utilities can be made available in every country. Per capita consumption of electricity, water and other utility services is quite high in industrialized countries. While production in mining and quarrying is decreasing in industrialized countries and population growth - which determines demand for utilities - is moderate, the share of industrialized countries in the world total output of mining and utilities is gradually declining. Industrialized countries accounted for nearly 70 per cent of the world total output of mining and utilities in 1990. The share of mining and utilities in developing and emerging industrial economies increased by up to 50 per cent in 2015. Demand for energy products such as electricity, natural gas and coal in developing and emerging industrial economies is rising to foster the manufacturing sector's high growth rate.

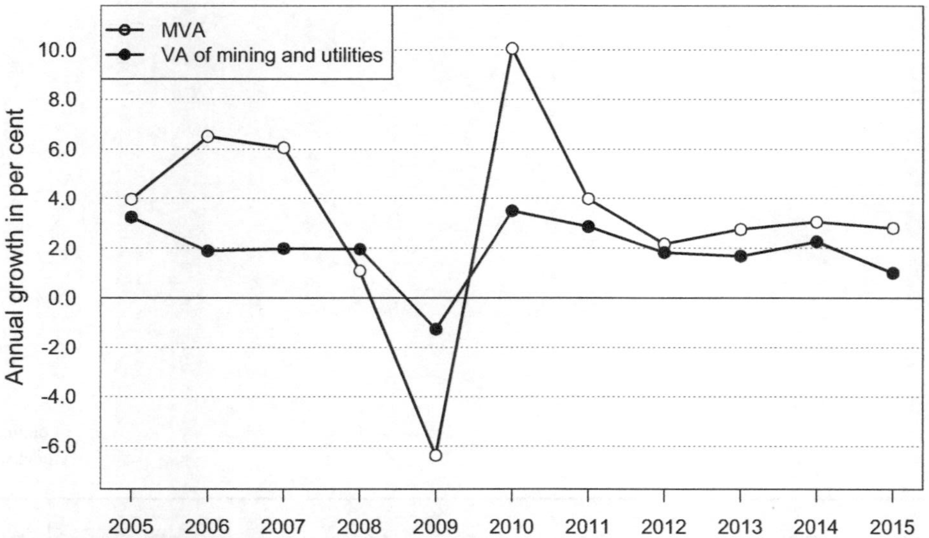

Source: UNIDO Database

Figure 4: Annual growth of global value added of mining and utilities and of manufacturing at constant 2010 USD

A large share of goods and services produced by the utilities sector constitute an essential part of household consumption, including electricity, cooking gas and drinking water. The production process in mining and utilities involves a network of mines, structures, pipelines, electricity mains, etc. While demand for essential utility products and services remains relatively stable, any interruption in production is quite costly for producers due to the infrastructure involved. The production process in the mining and utilities sectors is therefore less exposed to external volatility. During the years of the global financial crisis fluctuations in manufacturing production were much higher than in the mining and utilities sectors. At the same time, the growth of mining and utilities followed the same trend as that of manufacturing. This is primarily attributable to the fact that the mining and manufacturing sectors are closely related and that the majority of products from mining and quarrying activities are further processed by manufacturing before reaching the final consumer. A brief period of growth of the global manufacturing sector for 2012-2014 provided favorable conditions for the growth of the mining and utilities sectors. However, world value added of manufacturing and mining and utilities sectors began declining again, primarily due to consistently falling oil prices.

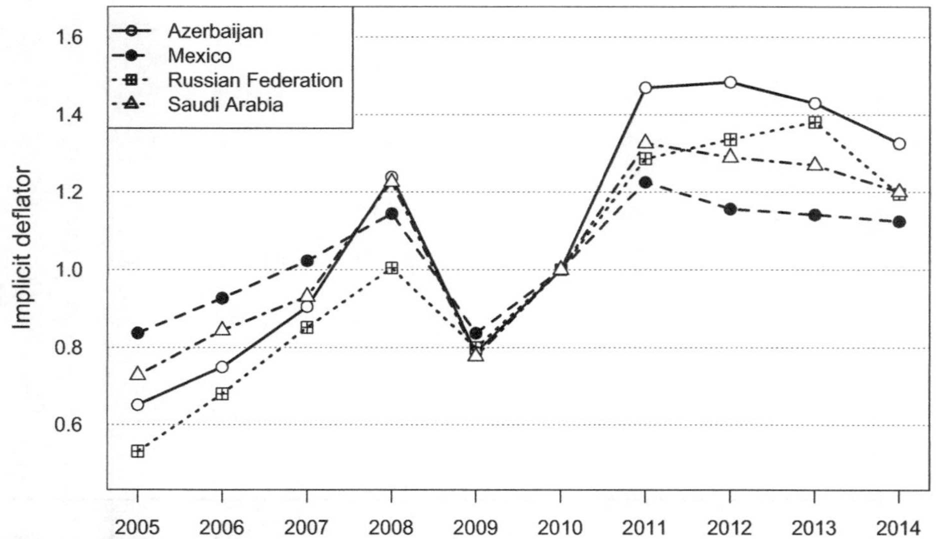

Source: UNIDO Database

Figure 5: Relative change in price of mining and utility products (implicit deflator) in selected countries

As previously mentioned, constant price value added figures have been rebased in the current publication from 2005 to 2010 to reflect more recent price levels. In general, such rebasing raises the value of commodities to higher levels proportionate to changes in price. However, the price changes diverged in the different countries. A comparison of selected countries in the graph above reveals that the price change in Azerbaijan was higher than in Mexico. Average prices fell in all countries in 2009 due to the severe drop in demand at the height of the financial crisis. Prices increased in different directions in the years of recovery. Recently, average prices of mining and utility products have again been gradually declining on account of the falling oil prices.

According OPEC data, the basket price of oil has dropped from a record high of 109.45 USD per barrel in 2012 to USD 30.6 in early 2016. The average price of crude oil per barrel was USD 49.49 in 2015. A significant decrease in oil prices has pushed down the global value added of the mining and utilities sectors, and even more so in major oil producing countries.

Table 1.1

DISTRIBUTION OF WORLD VALUE ADDED OF MINING AND UTILITIES SECTORS, 2005-2015

Year	Industrialized Economies					Developing and Emerging Industrial Economies			
	Europe	East Asia	West Asia	North America	Others	Africa	Asia & Pacific	Europe	Latin America
	Percentage share in world total value added (at constant 2010 prices)								
2005	21.7	4.9	5.5	17.9	3.4	6.9	24.7	2.9	12.1
2006	20.4	5.1	5.8	18.2	3.5	6.8	25.3	2.7	12.2
2007	19.9	4.9	5.6	18.5	3.6	6.8	26.0	2.6	12.1
2008	19.5	5.1	5.6	18.4	3.7	6.6	26.5	2.5	12.1
2009	19.6	4.3	5.3	19.0	3.9	6.5	26.9	2.6	11.9
2010	18.9	4.6	5.6	18.5	4.0	6.4	27.5	2.5	12.0
2011	17.5	4.0	6.0	18.6	4.0	6.3	29.1	2.6	11.9
2012	17.7	3.4	6.3	19.2	4.1	6.0	28.7	2.5	12.1
2013	17.0	3.5	6.3	19.1	4.2	5.6	29.6	2.5	12.2
2014 a/	16.5	3.7	6.2	19.0	4.2	5.4	30.7	2.3	12.0
2015 b/	16.1	3.7	6.2	19.2	4.2	5.4	31.3	2.2	11.7

a/ Provisional.
b/ Estimate.

Table 1.2 DISTRIBUTION OF WORLD VALUE ADDED OF MINING AND UTILITIES SECTORS AND POPULATION, SELECTED YEARS

(Percentage)

Country group	Value added at constant 2010 prices					Population				
	2005	2010	2013	2014 a/	2015 b/	2005	2010	2013	2014	2015
Africa	6.9	6.5	5.6	5.4	5.3	11.9	12.7	13.2	13.4	13.6
Asia & Pacific	38.4	41.6	43.6	44.7	45.4	61.2	61.0	60.8	60.8	60.7
Europe	24.6	21.4	19.5	18.8	18.4	12.8	12.2	11.9	11.8	11.7
Latin America	12.2	12.0	12.2	12.1	11.7	8.9	8.9	8.9	8.9	8.9
North America	17.9	18.5	19.1	19.0	19.2	5.2	5.2	5.2	5.1	5.1
World	100.0	100.0	100.0	100.0	100.0	100.0	100.0	100.0	100.0	100.0

a/ Provisional.
b/ Estimate.

Table 1.3 **GROWTH OF TOTAL AND PER-CAPITA VALUE ADDED OF MINING AND UTILITIES SECTORS, AT CONSTANT 2010 PRICES**

Country groups	Total value added		Per-capita value added	
	2005-2010	2010-2015	2005-2010	2010-2015
Industrialized Economies	0.8	1.1	0.1	0.7
EU [a/]	-0.7	-1.5	-1.1	-1.8
Other Europe	-1.4	-0.2	-1.5	-0.1
East Asia	-0.8	-2.1	-1.2	-2.5
West Asia	0.9	3.7	-10.3	0.7
North America	2.3	2.6	1.3	1.7
Others	4.5	3.5	2.8	2.3
Developing & Emerging Industrial Economies	2.2	2.7	0.9	1.4
Africa	-0.1	-2.1	-2.5	-4.4
Asia & Pacific	3.6	4.3	2.5	3.3
Europe	-1.4	-0.8	-1.6	-1.0
Latin America	1.1	1.7	-0.1	0.6
World	1.4	1.9	0.3	0.8

a/ Excluding non-industrialized EU economies.

Table 1.4

SHARE OF VALUE ADDED OF MINING AND UTILITIES SECTORS IN GDP, SELECTED YEARS
(Percentage)

Country groups	At constant 2010 prices					
	2005	2010	2012	2013	2014 [b]	2015 [c]
Industrialized Economies	4.9	4.9	4.9	4.8	4.8	4.7
EU [a]	3.6	3.3	3.1	3.0	2.9	2.8
Other Europe	13.8	11.3	10.9	10.4	10.6	10.4
East Asia	2.7	2.5	1.9	2.0	2.0	2.1
West Asia	49.0	43.0	44.7	43.3	42.4	41.6
North America	4.4	4.8	4.9	4.9	4.9	4.8
Others	8.4	9.3	9.7	9.8	10.3	10.3
Developing & Emerging Industrial Economies	12.6	10.3	10.0	9.8	9.7	9.5
Africa	20.6	16.2	14.6	13.4	12.7	12.3
Asia & Pacific	13.6	10.7	10.3	10.1	10.2	9.8
Europe	6.1	5.0	5.0	4.9	4.6	4.5
Latin America	11.2	9.8	9.7	9.7	9.8	9.7
World	6.9	6.5	6.5	6.4	6.4	6.4

a/ Excluding non-industrialized EU economies.
b/ Provisional.
c/ Estimate.

Part II
COUNTRY TABLES

Afghanistan

ISIC Revision 3		Number of establishments					Number of employees					Wages and salaries paid to employees				
			(number)					(number)					(Afghan Afghani)			
ISIC	Industry	Note	2010	2011	2012	2013	Note	2010	2011	2012	2013	Note	2010	2011	2012	2013
10	Mining of coal and lignite; extraction of peat		...	...	...	...		...	...	...	...		...	...	...	...
101	Mining and agglomeration of hard coal		7	7	...	...		...	...	...	...		...	...	...	...
102	Mining and agglomeration of lignite		...	...	...	...		...	...	...	...		...	...	...	...
103	Extraction and agglomeration of peat		...	...	15	15		...	...	...	...		...	...	...	...
11	Extraction of crude oil and natural gas, etc.		...	...	...	...		...	...	...	...		...	...	...	...
111	Extraction of crude oil and natural gas		1	1	3	3		...	...	...	...		...	...	...	...
112	Oil, gas service activities		...	...	...	...		...	...	...	...		...	...	...	...
12	Mining of uranium and thorium ores		...	...	...	...		...	...	...	...		...	...	...	...
120	Mining of uranium and thorium ores		...	...	...	...		...	...	...	...		...	...	...	...
13	Mining of metal ores		...	...	...	...		...	...	...	...		...	...	...	...
131	Iron ores		...	...	...	...		...	...	...	...		...	...	...	...
132	Non-ferrous metal ores,excl.uranium,thorium		...	...	...	...		...	...	...	...		...	...	...	...
14	Other mining and quarrying		...	...	26	26		...	...	...	...		...	...	...	...
141	Quarrying of stone, sand and clay		...	...	9	9		...	...	...	...		...	...	...	...
142	Mining and quarrying n.e.c.		...	...	17	17		...	...	...	...		...	...	...	...
C	Total mining and quarrying	a/	8	8	44	44		...	...	...	...		...	...	...	...
40	Electricity, gas, steam and hot water supply		...	...	...	...		...	...	...	...		...	...	...	...
401	Electricity production, collection, distribution		27	27	27	27		...	...	...	...		...	...	...	...
402	Gas production and distribution		1	1	3	3		...	...	...	...		...	...	...	...
403	Steam and hot water supply		...	...	...	...		...	...	...	...		...	...	...	...
41	Water collection, purification, distribution		...	...	...	...		...	...	...	...		...	...	...	...
410	Water collection, purification, distribution		...	...	...	...		...	...	...	...		...	...	...	...
E	Total electricity, gas and water supply	a/	28	28	30	30		...	...	...	...		...	...	...	...

a/ Sum of available data.

Afghanistan

ISIC Revision 3		Output (valuation not defined)					Value added				
			(millions of Afghan Afghani)					(millions of Afghan Afghani)			
ISIC	Industry	Note	2010	2011	2012	2013	Note	2010	2011	2012	2013
10	Mining of coal and lignite; extraction of peat		...	...	...	...		...	...	...	...
101	Mining and agglomeration of hard coal		1035.7	867.9	...	...		...	...	...	...
102	Mining and agglomeration of lignite		...	...	...	...		...	...	...	...
103	Extraction and agglomeration of peat		...	...	943.0	1062.0		...	...	...	...
11	Extraction of crude oil and natural gas, etc.		...	...	...	...		...	...	...	...
111	Extraction of crude oil and natural gas		258.6	256.9	248.0	227.0		...	...	...	...
112	Oil, gas service activities		...	...	...	...		...	...	...	...
12	Mining of uranium and thorium ores		...	...	...	...		...	...	...	...
120	Mining of uranium and thorium ores		...	...	...	...		...	...	...	...
13	Mining of metal ores		...	...	...	...		...	...	...	...
131	Iron ores		...	...	...	...		...	...	...	...
132	Non-ferrous metal ores,excl.uranium,thorium		...	...	...	...		...	...	...	...
14	Other mining and quarrying		...	...	...	553.0		...	...	...	...
141	Quarrying of stone, sand and clay		...	...	...	35.0		...	...	...	...
142	Mining and quarrying n.e.c.		...	...	107.0	518.0		...	...	...	...
C	Total mining and quarrying	a/	1294.3	1124.8	1298.0	1842.0		...	...	...	...
40	Electricity, gas, steam and hot water supply		...	...	...	...		...	...	...	...
401	Electricity production, collection, distribution		1046.9	1071.2	1022.0	1049.0		...	...	...	...
402	Gas production and distribution		258.6	256.9	248.0	227.0		...	...	...	...
403	Steam and hot water supply		...	...	...	...		...	...	...	...
41	Water collection, purification, distribution		...	...	...	...		...	...	...	...
410	Water collection, purification, distribution		...	...	...	...		...	...	...	...
E	Total electricity, gas and water supply	a/	1305.5	1328.1	1270.0	1276.0		...	...	...	...

a/ Sum of available data.

Albania

ISIC Revision 3		Number of enterprises					Number of employees					Wages and salaries paid to employees				
			(number)					(number)					(millions of Albanian Leks)			
ISIC	Industry	Note	2010	2011	2012	2013	Note	2010	2011	2012	2013	Note	2010	2011	2012	2013
10	Mining of coal and lignite; extraction of peat		3	9	7	5		34	14	19	10		11.6	4.6	7.5	2.1
101	Mining and agglomeration of hard coal	a/	...	...	...	...	a/	...	...	...	...	a/	...	...	...	...
102	Mining and agglomeration of lignite	a/	3	9	7	5	a/	34	14	19	10	a/	11.6	4.6	7.5	2.1
103	Extraction and agglomeration of peat	a/	...	...	...	...	a/	...	...	...	...	a/	...	...	...	...
11	Extraction of crude oil and natural gas, etc.		19	22	23	22		4231	4233	4885	5307		4724.6	5722.0	7085.1	7981.4
111	Extraction of crude oil and natural gas	b/	19	22	23	22	b/	4231	4233	4885	5307	b/	4724.6	5722.0	7085.1	7981.4
112	Oil, gas service activities	b/	...	...	...	...	b/	...	...	...	...	b/	...	...	...	...
12	Mining of uranium and thorium ores		...	...	...	...		...	...	...	...		...	...	...	...
120	Mining of uranium and thorium ores		...	...	...	...		...	...	...	...		...	...	...	...
13	Mining of metal ores		145	218	156	221		1547	3148	3106	3410		757.6	1825.6	2116.4	2438.1
131	Iron ores	c/	145	218	156	221	c/	1547	3148	3106	3410	c/	757.6	1825.6	2116.4	2438.1
132	Non-ferrous metal ores,excl.uranium,thorium	c/	...	...	...	...	c/	...	...	...	...	c/	...	...	...	...
14	Other mining and quarrying		242	212	327	280		1580	1314	1494	1629		569.9	507.9	563.9	611.6
141	Quarrying of stone, sand and clay	d/	242	212	327	280	d/	1580	1314	1494	1629	d/	569.9	507.9	563.9	611.6
142	Mining and quarrying n.e.c.	d/	...	...	...	...	d/	...	...	...	...	d/	...	...	...	...
C	Total mining and quarrying	e/	409	461	513	528	e/	7392	8709	9504	10356	e/	6063.7	8060.2	9772.9	11033.3
40	Electricity, gas, steam and hot water supply		114	163	160	265		7680	7233	6890	7223		5257.8	4397.2	5212.7	5072.6
401	Electricity production, collection, distribution	f/	114	163	160	265	f/	7680	7233	6890	7223	f/	5257.8	4397.2	5212.7	5072.6
402	Gas production and distribution	f/	...	...	...	...	f/	...	...	...	...	f/	...	...	...	...
403	Steam and hot water supply	f/	...	...	...	...	f/	...	...	...	...	f/	...	...	...	...
41	Water collection, purification, distribution		56	61	61	60		5102	5206	5513	5715		1786.6	1854.6	2153.9	2271.2
410	Water collection, purification, distribution		56	61	61	60		5102	5206	5513	5715		1786.6	1854.6	2153.9	2271.2
E	Total electricity, gas and water supply	e/	170	224	221	325	e/	12782	12439	12403	12938	e/	7044.4	6251.9	7366.6	7343.8

a/ 102 includes 101 and 103.
b/ 111 includes 112.
c/ 131 includes 132.
d/ 141 includes 142.
e/ Sum of available data.
f/ 401 includes 402 and 403.

Albania

ISIC Revision 3		Output (valuation not defined) a/					Value added (valuation not defined) b/				
			(millions of Albanian Leks)					(millions of Albanian Leks)			
ISIC	Industry	Note	2010	2011	2012	2013	Note	2010	2011	2012	2013
10	Mining of coal and lignite; extraction of peat		40.1	17.2	17.8	18.9		16.7	17.0	14.8	7.1
101	Mining and agglomeration of hard coal	c/	...	...	...	...	c/	...	...	...	...
102	Mining and agglomeration of lignite	c/	40.1	17.2	17.8	18.9	c/	16.7	17.0	14.8	7.1
103	Extraction and agglomeration of peat	c/	...	...	...	...	c/	...	...	...	...
11	Extraction of crude oil and natural gas, etc.		24219.8	41137.6	52818.4	67492.4		16156.3	33522.8	38542.2	46653.9
111	Extraction of crude oil and natural gas	d/	24219.8	41137.6	52818.4	67492.4	d/	16156.3	33522.8	38542.2	46653.9
112	Oil, gas service activities	d/	...	...	...	...	d/	...	...	...	...
12	Mining of uranium and thorium ores		...	...	...	...		...	...	...	...
120	Mining of uranium and thorium ores		...	...	...	...		...	...	...	...
13	Mining of metal ores		9762.1	14102.8	15947.0	15925.6		4756.5	4665.3	8929.0	8455.8
131	Iron ores	e/	9762.1	14102.8	15947.0	15925.6	e/	4756.5	4665.3	8929.0	8455.8
132	Non-ferrous metal ores,excl.uranium,thorium	e/	...	...	...	...	e/	...	...	...	...
14	Other mining and quarrying		6106.7	6819.9	6619.9	8041.6		1658.4	1565.6	2089.7	1655.2
141	Quarrying of stone, sand and clay	f/	6106.7	6819.9	6619.9	8041.6	f/	1658.4	1565.6	2089.7	1655.2
142	Mining and quarrying n.e.c.	f/	...	...	...	...	f/	...	...	...	...
C	Total mining and quarrying	g/	40128.7	62077.5	75403.0	91478.6	g/	22587.9	39770.7	49575.8	56772.0
40	Electricity, gas, steam and hot water supply		74463.3	63919.2	76939.6	89526.3		45177.8	10480.5	13510.3	25100.7
401	Electricity production, collection, distribution	h/	74463.3	63919.2	76939.6	89526.3	h/	45177.8	10480.5	13510.3	25100.7
402	Gas production and distribution	h/	...	...	...	...	h/	...	...	...	...
403	Steam and hot water supply	h/	...	...	...	...	h/	...	...	...	...
41	Water collection, purification, distribution		6002.4	5902.2	7822.8	7214.2		3649.7	2181.2	4954.8	4249.4
410	Water collection, purification, distribution		6002.4	5902.2	7822.8	7214.2		3649.7	2181.2	4954.8	4249.4
E	Total electricity, gas and water supply	g/	80465.7	69821.4	84762.5	96740.5	g/	48827.5	12661.7	18465.1	29350.1

a/ Output is at basic prices.
b/ Value added is at basic prices.
c/ 102 includes 101 and 103.
d/ 111 includes 112.
e/ 131 includes 132.
f/ 141 includes 142.
g/ Sum of available data.
h/ 401 includes 402 and 403.

Armenia

ISIC Revision 4		Number of establishments					Number of employees					Wages and salaries paid to employees				
			(number)					(number)					(millions of Armenian Drams)			
ISIC	Industry	Note	2010	2011	2012	2013	Note	2010	2011	2012	2013	Note	2010	2011	2012	2013
051	Mining of hard coal		-	-	...	...		...	...	...	...		...	...	...	...
052	Mining of lignite		-	-	...	...		...	...	...	...		...	...	...	...
061	Extraction of crude petroleum		-	-	...	...		...	...	...	...		...	...	...	...
062	Extraction of natural gas		-	-	...	...		...	...	...	...		...	...	...	...
071	Mining of iron ores		-	-	...	...		...	...	...	...		...	...	...	...
072	Mining of non-ferrous metal ores		7	7	8	9		...	...	...	...		...	...	...	...
081	Quarrying of stone, sand and clay		44	47	51	55		...	...	...	...		...	...	...	...
089	Mining and quarrying n.e.c.		11	11	12	11		...	...	...	...		...	...	...	...
091	Support activities for petroleum		-	-	...	...		...	...	...	...		...	...	...	...
099	Support activities for other mining, quarrying		-	-	...	...		...	...	...	...		...	...	...	...
B	Mining and quarrying		62	65	71	75		10700	15300	7479	8146		27848	43637	25819	30992
351	Electric power generation, transmission		102	121	142	162		...	...	...	...		...	...	...	...
352	Manufacture of gas		1	1	1	1		...	...	...	...		...	...	...	...
353	Steam and air conditioning supply		2	2	1	2		...	...	...	...		...	...	...	...
D	Electricity, gas, steam & air conditioning		105	124	144	165		31600	25700	20256	20295		56796	46936	43377	47102
360	Water collection, treatment and supply		2	2	3	4		...	...	...	...		...	...	...	...
370	Sewerage		9	9	9	7		...	...	...	...		...	...	...	...
381	Waste collection		60	63	66	66		...	...	...	...		...	...	...	...
382	Waste treatment and disposal		3	3	3	3		...	...	...	...		...	...	...	...
383	Materials recovery		2	2	1	1		...	...	...	...		...	...	...	...
390	Remediation activities		-	-	...	...		...	...	...	...		...	...	...	...
E	Water supply;sewerage,waste management		76	79	82	81		9800	8100	5109	4712		11779	10789	8132	7896

Armenia

ISIC Revision 4		Output at producers' prices					Value added (valuation not defined)				
		Note	(millions of Armenian Drams)				Note	(millions of Armenian Drams)			
ISIC	Industry		2010	2011	2012	2013		2010	2011	2012	2013
051	Mining of hard coal		-	-	...	...		...	...	...	...
052	Mining of lignite		-	-	...	...		...	...	...	...
061	Extraction of crude petroleum		-	-	...	...		...	...	...	...
062	Extraction of natural gas		-	-	...	...		...	...	...	...
071	Mining of iron ores		-	-	...	...		...	...	...	...
072	Mining of non-ferrous metal ores		140446	165595	187426	191339		...	...	...	...
081	Quarrying of stone, sand and clay		3905	3700	4561	4946		...	...	...	...
089	Mining and quarrying n.e.c.		1186	874	1339	494		...	...	...	...
091	Support activities for petroleum		-	-	...	...		...	...	...	...
099	Support activities for other mining, quarrying		-	-	...	...		...	...	...	...
B	Mining and quarrying		145537	170169	193326	196779		89251	102846	...	...
351	Electric power generation, transmission		93173	143217	182789	214478		...	...	...	...
352	Manufacture of gas		27959	22433	28932	36013		...	...	...	...
353	Steam and air conditioning supply		25	33	3	12		...	...	...	...
D	Electricity, gas, steam & air conditioning		121157	165683	211724	250503		97513	131188	...	...
360	Water collection, treatment and supply		10614	10857	11379	12020		...	...	...	...
370	Sewerage		2273	2312	2682	2089		...	...	...	...
381	Waste collection		4839	4969	3242	3586		...	...	...	...
382	Waste treatment and disposal		51	60	85	76		...	...	...	...
383	Materials recovery		157	139	50	...		...	...	...	...
390	Remediation activities		-	-	...	...		...	...	...	...
E	Water supply;sewerage,waste management		17934	18337	17438	17771		15854	14500	...	...

Australia

ISIC Revision 4		Number of enterprises					Number of employees					Wages and salaries paid to employees				
			(number)					(number)					(millions of Australian Dollars)			
ISIC	Industry	Note	2010	2011	2012	2013	Note	2010	2011	2012	2013	Note	2010	2011	2012	2013
051	Mining of hard coal		681	698	643	...	a/	34000	38000	45000	43000	a/	4119	4996	6052	6247
052	Mining of lignite		141	161	148	...	a/	...	...	...	...	a/	...	...	...	...
061	Extraction of crude petroleum		130	134	129	...	b/	15000	15000	18000	19000	b/	2458	2594	3401	3737
062	Extraction of natural gas		130	134	129	...	b/	...	...	...	...	b/	...	...	...	...
071	Mining of iron ores		118	117	117	...	c/	45000	50000	69000	69000	c/	5396	6302	8208	9835
072	Mining of non-ferrous metal ores		1009	1054	993	...	c/	...	...	...	...	c/	...	...	...	...
081	Quarrying of stone, sand and clay		1369	1360	1286	...	d/	13000	13000	13000	13000	d/	897	939	1018	989
089	Mining and quarrying n.e.c.		729	734	661	...	d/	...	...	...	...	d/	...	...	...	...
091	Support activities for petroleum		2182	2254	2092	...	e/	37000	43000	49000	46000	e/	3763	3912	4814	4911
099	Support activities for other mining, quarrying		3373	3599	3373	...	e/	...	...	...	...	e/	...	...	...	...
B	Mining and quarrying		9862	10245	9571	...		144000	159000	194000	190000		16635	18743	23493	25720
351	Electric power generation, transmission		477	494	484	...		50000	51000	53000	52000		5231	5292	5833	5854
352	Manufacture of gas		295	290	268	...		2000	2000	3000	3000		166	188	222	237
353	Steam and air conditioning supply		237	231	219	...		...	...	...	...		...	...	...	...
D	Electricity, gas, steam & air conditioning		1009	1015	971	...		52000	53000	56000	55000		5397	5480	6055	6091
360	Water collection, treatment and supply		237	231	219	...		...	29000f/	27000f/	26000f/		...	2250f/	2231f/	2200f/
370	Sewerage		427	427	403	...		...	...f/	...f/	...f/		...	...f/	...f/	...f/
381	Waste collection		2718	2730	2560	...	g/	29000	27000	31000	35000	g/	1629	1777	2007	2228
382	Waste treatment and disposal		995	1018	975	...	g/	...	...	...	...	g/	...	...	...	...
383	Materials recovery		288	288	269	...	g/	...	...	...	...	g/	...	...	...	...
390	Remediation activities		209	212	198	...		...	...	...	...		...	...	...	...
E	Water supply;sewerage,waste management		4874	4906	4624	...		56000	56000	58000	61000		3553	4027	4238	4428

a/ 051 includes 052.
b/ 061 includes 062.
c/ 071 includes 072.
d/ 081 includes 089.
e/ 091 includes 099.
f/ 360 includes 370.
g/ 381 includes 382 and 383.

Australia

ISIC Revision 4		Output (valuation not defined)					Value added (valuation not defined)				
			(millions of Australian Dollars)					(millions of Australian Dollars)			
ISIC	Industry	Note	2010	2011	2012	2013	Note	2010	2011	2012	2013
051	Mining of hard coal	a/	45045	54715	58417	48518	a/	22785	27111	25535	16909
052	Mining of lignite	a/	...	...	...	...	a/	...	...	...	...
061	Extraction of crude petroleum	b/	27637	33947	35196	38822	b/	21893	27080	27847	29682
062	Extraction of natural gas	b/	...	...	...	...	b/	...	...	...	...
071	Mining of iron ores	c/	59586	89100	101843	92821	c/	36356	63922	70371	58862
072	Mining of non-ferrous metal ores	c/	...	...	...	...	c/	...	...	...	...
081	Quarrying of stone, sand and clay	d/	5296	6134	6038	5797	d/	2006	2682	2079	2625
089	Mining and quarrying n.e.c.	d/	...	...	...	...	d/	...	...	...	...
091	Support activities for petroleum	e/	13742	13778	15936	15189	e/	5420	5482	6732	6366
099	Support activities for other mining, quarrying	e/	...	...	...	...	e/	...	...	...	...
B	Mining and quarrying		151305	197673	217430	201147		88461	126276	132563	114443
351	Electric power generation, transmission		56388	59752	63985	73917		20444	21704	24201	25985
352	Manufacture of gas		7438	...	8570	9623		982	1422	1801	1994
353	Steam and air conditioning supply		...	...	...	...		...	...	...	...
D	Electricity, gas, steam & air conditioning		63826	...	72555	83540		21426	23126	26002	27979
360	Water collection, treatment and supply		...	15582f/	17558f/	18207f/		...	9046f/	10001f/	10261f/
370	Sewerage		...	...f/	...f/	...f/		...	...f/	...f/	...f/
381	Waste collection		9383g/	...	11798g/	12099g/	g/	3316	3512	4078	4264
382	Waste treatment and disposal		...g/	...	...g/	...g/	g/	...	...	...	...
383	Materials recovery		...g/	...	...g/	...g/	g/	...	...	...	...
390	Remediation activities		...	...	...	...		...	...	...	...
E	Water supply;sewerage,waste management		22406	...	29356	30306		10676	12558	14079	14525

a/ 051 includes 052.
b/ 061 includes 062.
c/ 071 includes 072.
d/ 081 includes 089.
e/ 091 includes 099.
f/ 360 includes 370.
g/ 381 includes 382 and 383.

Austria

ISIC Revision 4		Number of enterprises					Number of employees					Wages and salaries paid to employees				
			(number)					(number)					(millions of Euros)			
ISIC	Industry	Note	2010	2011	2012	2013	Note	2010	2011	2012	2013	Note	2010	2011	2012	2013
05	Mining of coal and lignite		-	-	-	-		-	-	-	-		-	-	-	-
051	Mining of hard coal		-	-	-	-		-	-	-	-		-	-	-	-
052	Mining of lignite		-	-	-	-		-	-	-	-		-	-	-	-
06	Extraction of crude petroleum & natural gas		2	2	2	2		...	...	...	...		...	...	...	...
061	Extraction of crude petroleum		2	2	2	2		...	...	...	...		...	...	...	...
062	Extraction of natural gas		-	-	-	-		-	-	-	-		-	-	-	-
07	Mining of metal ores		2	2	2	2		...	...	...	...		...	...	...	...
071	Mining of iron ores		2	2	2	2		...	...	...	...		...	...	...	...
072	Mining of non-ferrous metal ores		-	-	-	-		-	-	-	-		-	-	-	-
08	Other mining and quarrying		343	341	342	334		4656	4507	4567	4543		170	170	176	177
081	Quarrying of stone, sand and clay		323	320	322	315		4019	3846	3917	3893		141	141	146	147
089	Mining and quarrying n.e.c.		20	21	20	19		637	661	650	650		29	29	29	31
09	Mining support service activities		11	10	9	15		42	41	38	117		3	2	2	5
091	Support activities for petroleum		8	5	4	8		24	22	16	86		2	1	1	4
099	Support activities for other mining, quarrying		3	5	5	7		18	19	22	31		1	1	1	1
B	Mining and quarrying		358	355	355	353		5933	5816	5874	5919		265	268	274	280
35	Electricity, gas, steam & air conditioning		1878	1961	2142	2256		27402	27397	27519	27674		1518	1538	1621	1642
351	Electric power generation, transmission		995	1031	1165	1259		21957	21827	21931	24352		1255	1274	1355	1482
352	Manufacture of gas		38	40	44	41		2997	3073	2983	1808		152	153	150	107
353	Steam and air conditioning supply		845	890	933	956		2448	2497	2605	1514		111	112	115	54
D	Electricity, gas, steam & air conditioning		1878	1961	2142	2256		27402	27397	27519	27674		1518	1538	1621	1642
36	Water collection, treatment and supply		571	578	566	586		2356	2336	2347	1762		91	92	95	71
37	Sewerage		695	673	659	675		2483	3132	3195	3200		84	110	116	119
38	Waste collection,treatment,disposal activities		788	749	773	775		12572	12760	13269	14371		414	428	455	508
381	Waste collection		326	307	317	323		8616	8618	8901	9834		276	281	296	339
382	Waste treatment and disposal		317	304	314	315		2039	2099	2226	2281		73	78	84	89
383	Materials recovery		145	138	142	137		1917	2043	2142	2256		65	70	75	80
39	Remediation activities		6	8	14	14		107	104	100	87		3	3	3	3
E	Water supply;sewerage,waste management		2060	2008	2012	2050		17518	18332	18911	19420		592	633	669	702

Austria

ISIC Revision 4		Output (valuation not defined)					Value added at factor values				
			(millions of Euros)					(millions of Euros)			
ISIC	Industry	Note	2010	2011	2012	2013	Note	2010	2011	2012	2013
05	Mining of coal and lignite		-	-	-	-		-	-	-	-
051	Mining of hard coal		-	-	-	-		-	-	-	-
052	Mining of lignite		-	-	-	-		-	-	-	-
06	Extraction of crude petroleum & natural gas		...	...	...	...		...	...	...	...
061	Extraction of crude petroleum		...	...	...	...		...	...	...	...
062	Extraction of natural gas		-	-	-	-		-	-	-	-
07	Mining of metal ores		...	...	...	...		...	...	...	...
071	Mining of iron ores		...	...	...	...		...	...	...	...
072	Mining of non-ferrous metal ores		-	-	-	-		-	-	-	-
08	Other mining and quarrying		995	1006	1034	1083		397	396	403	450
081	Quarrying of stone, sand and clay		801	792	817	858		326	321	319	354
089	Mining and quarrying n.e.c.		194	214	217	225		71	75	84	97
09	Mining support service activities		14	13	17	36		3	4	3	10
091	Support activities for petroleum		12	9	13	32		3	3	2	8
099	Support activities for other mining, quarrying		2	4	4	5		-	1	1	2
B	Mining and quarrying		2185	2534	2607	2501		1026	1295	1294	1245
35	Electricity, gas, steam & air conditioning		29456	33237	38680	38780		5217	5603	5472	5591
351	Electric power generation, transmission		22757	23735	24773	26214		4337	4325	4663	5125
352	Manufacture of gas		5464	8216	12446	11544		451	838	349	167
353	Steam and air conditioning supply		1235	1286	1461	1022		428	440	460	299
D	Electricity, gas, steam & air conditioning		29456	33237	38680	38780		5217	5603	5472	5591
36	Water collection, treatment and supply		485	494	564	470		296	313	363	310
37	Sewerage		547	751	782	797		352	418	443	444
38	Waste collection,treatment,disposal activities		2645	2795	2972	3300		943	990	1034	1115
381	Waste collection		1347	1350	1397	1669		553	584	603	712
382	Waste treatment and disposal		594	542	550	554		221	234	248	253
383	Materials recovery		704	903	1025	1077		169	172	183	151
39	Remediation activities		11	13	37	55		5	5	5	8
E	Water supply;sewerage,waste management		3689	4053	4355	4622		1597	1725	1845	1877

Azerbaijan

ISIC Revision 4		Number of establishments					Number of employees					Wages and salaries paid to employees				
			(number)					(number)					(thousands of Azerbaijani Manat)			
ISIC	Industry	Note	2010	2011	2012	2013	Note	2010	2011	2012	2013	Note	2010	2011	2012	2013
051	Mining of hard coal		-	-	...	...		-	-	...	...		-	-	...	...
052	Mining of lignite		-	-	...	...		-	-	...	...		-	-	...	...
061	Extraction of crude petroleum		22	24	25	26		20055	20241	20649	21001		287876	330673	404468	404468
062	Extraction of natural gas		5	4	3	3		4681	3247	3050	2959		30413	34037	42034	42034
071	Mining of iron ores		2	2	2	2		165	305	362	393		574	1204	1511	1511
072	Mining of non-ferrous metal ores		1	1	1	1		418	487	525	554		4814	5778	6085	6086
081	Quarrying of stone, sand and clay		219	195	191	198		2748	2742	2951	5452		14267	10297	8646	8646
089	Mining and quarrying n.e.c.		8	9	9	10		705	7960	850	860		2193	2523	3412	3413
091	Support activities for petroleum		54	54	58	78		7788	7829	8434	8450		94710	116520	144671	144671
099	Support activities for other mining, quarrying		-	-	...	...		-	-	...	...		-	-	...	...
B	Mining and quarrying		311	289	289	318		36560	42811	36821	39669		434847	501032	610829	610829
351	Electric power generation, transmission		118	108	105	105		17807	17200	16810	16985		77279	88121	16810	87869
352	Manufacture of gas		72	72	74	73		8604	9899	10453	10631		35201	46072	10453	62033
353	Steam and air conditioning supply		36	36	36	42		2083	2190	2317	2896		5317	6014	78297	7132
D	Electricity, gas, steam & air conditioning		226	216	215	220		28494	29289	29580	30512		117797	140207	29580	157034
360	Water collection, treatment and supply		66	67	68	81		10169	9594	9450	11323		28146	32380	37459	37459
370	Sewerage		96	96	72	16		9041	9347	4448	774		16504	18854	4448	9199
381	Waste collection		-	1	15	61		-	1019	2670	5488		-	3436	2670	8455
382	Waste treatment and disposal		24	26	9	10		1807	1192	593	560		5309	3508	593	984
383	Materials recovery		7	7	7	8		152	166	172	163		692	786	172	762
390	Remediation activities		11	15	44	49		839	1153	6450	7165		1310	3113	6450	21434
E	Water supply;sewerage,waste management		204	212	215	225		22008	22471	23783	25473		51961	62077	78292	78293

Azerbaijan

ISIC Revision 4		Output at factor values					Value added at factor values				
			(thousands of Azerbaijani Manat)					(thousands of Azerbaijani Manat)			
ISIC	Industry	Note	2010	2011	2012	2013	Note	2010	2011	2012	2013
051	Mining of hard coal		-	-	...	...		-	-	...	...
052	Mining of lignite		-	-	...	...		-	-	...	...
061	Extraction of crude petroleum		20154084	24927063	24668206	23524973		18215901	23688852	23138473	22299361
062	Extraction of natural gas		...	1128299	78758	132849		1020400	927664	68379	120095
071	Mining of iron ores		111	5680	4361	3515		47	2454	1295	1651
072	Mining of non-ferrous metal ores		62387	68637	63770	47726		23403	40076	21448	15574
081	Quarrying of stone, sand and clay		50933	72186	48610	46844		31628	20946	15388	15221
089	Mining and quarrying n.e.c.		8026	11062	17779	18430		3094	3450	5591	7168
091	Support activities for petroleum		542353	681419	725735	880997		187623	296806	319623	331105
099	Support activities for other mining, quarrying		-	-	...	...		-	-	...	...
B	Mining and quarrying		20862474	26894346	25607219	24655334		19482096	24980248	23570197	22790175
351	Electric power generation, transmission		612724	721022	722111	726150		148457	402701	432311	443265
352	Manufacture of gas		603387	820820	983516	1042155		252403	482224	640820	660441
353	Steam and air conditioning supply		9425	14004	18716	5554		1186	2114	9116	3824
D	Electricity, gas, steam & air conditioning		1225536	1555846	1724343	1773859		402046	887040	1082247	1107530
360	Water collection, treatment and supply		94342	115006	112662	129346		23210	29701	25068	61164
370	Sewerage		29832	35993	14521	22752		10475	14409	6358	4236
381	Waste collection		-	35	2814	...		-	...	1619	2377
382	Waste treatment and disposal		27388	28792	18043	7910		11286	19746	4156	1683
383	Materials recovery		...	...	3185	905		45	11	2428	534
390	Remediation activities		2526	4418	50484	59833		1109	2660	26643	21083
E	Water supply;sewerage,waste management		154500	184318	201708	225129		46141	66543	66272	91077

Bangladesh

ISIC Revision 4		Output (valuation not defined) a/					Value added (valuation not defined) b/				
		Note	(millions of Bangladeshi Taka)				Note	(millions of Bangladeshi Taka)			
ISIC	Industry		2010	2011	2012	2013		2010	2011	2012	2013
051	Mining of hard coal		...	4611	...	...		...	823	...	...
052	Mining of lignite		...	...	...	...		...	...	...	...
061	Extraction of crude petroleum		...	...	...	...		...	...	...	...
062	Extraction of natural gas		...	49588	...	...		...	47293	...	...
071	Mining of iron ores		...	...	...	...		...	...	...	...
072	Mining of non-ferrous metal ores		...	...	...	...		...	...	...	...
081	Quarrying of stone, sand and clay		...	52123	...	...		...	46480	...	...
089	Mining and quarrying n.e.c.		...	...	...	...		...	...	...	...
091	Support activities for petroleum		...	...	...	...		...	...	...	...
099	Support activities for other mining, quarrying		...	...	...	...		...	...	...	...
B	Mining and quarrying	c/	...	106322	...	...	c/	...	94596	...	...
351	Electric power generation, transmission		...	109238	...	...		...	80313	...	...
352	Manufacture of gas		...	10295	...	...		...	9664	...	...
353	Steam and air conditioning supply		...	...	...	...		...	...	...	...
D	Electricity, gas, steam & air conditioning		...	128185	...	...		...	95949	...	...
360	Water collection, treatment and supply		...	8652	...	...		...	5972	...	...
370	Sewerage		...	...	...	...		...	...	...	...
381	Waste collection		...	...	...	...		...	...	...	...
382	Waste treatment and disposal		...	...	...	...		...	...	...	...
383	Materials recovery		...	...	...	...		...	...	...	...
390	Remediation activities		...	...	...	...		...	...	...	...
E	Water supply;sewerage,waste management	c/	...	8652	...	...	c/	...	5972	...	...

a/ Output is at basic prices.
b/ Value added is at basic prices.
c/ Sum of available data.

Belarus

ISIC Revision 3		Number of establishments					Number of employees					Wages and salaries paid to employees a/				
			(number)					(thousands)					(billions of Belarusian Roubles)			
ISIC	Industry	Note	2010	2011	2012	2013	Note	2010	2011	2012	2013	Note	2010	2011	2012	2013
101	Mining and agglomeration of hard coal		...	...	...	...	b/	4.4	4.3	4.3	4.2	b/	59.9	91.7	166.9	221.0
102	Mining and agglomeration of lignite		...	...	...	...	b/	...	...	...	...	b/	...	...	...	...
103	Extraction and agglomeration of peat		...	...	...	...	b/	...	...	...	...	b/	...	...	...	...
111	Extraction of crude oil and natural gas		...	...	...	...	c/	3.9	4.2	4.4	4.4	c/	149.5	237.2	482.5	609.2
112	Oil, gas service activities		...	...	...	...	c/	...	...	...	...	c/	...	...	...	...
120	Mining of uranium and thorium ores		...	...	...	...		...	...	...	...		...	...	...	...
131	Iron ores		...	...	...	...	d/	7.4	7.6	7.3	7.6	d/	148.7	227.9	421.8	615.5
132	Non-ferrous metal ores,excl.uranium,thorium		...	...	...	...	d/	...	...	...	...	d/	...	...	...	...
141	Quarrying of stone, sand and clay		...	...	...	...		...	...	...	...		...	...	...	...
142	Mining and quarrying n.e.c.		...	...	...	...		...	...	...	...		...	...	...	...
C	Total mining and quarrying		44	44	46	50		15.7	16.1	15.9	16.2		358.0	556.8	1071.1	1445.7
401	Electricity production, collection, distribution		...	...	...	...	e/	101.1	98.5	98.1	97.1	e/	1707.7	2384.1	4507.6	6365.0
402	Gas production and distribution		...	...	...	...	e/	...	...	...	...	e/	...	...	...	...
403	Steam and hot water supply		...	...	...	...	e/	...	...	...	...	e/	...	...	...	...
410	Water collection, purification, distribution		...	...	...	...		15.7	15.4	14.7	14.5		220.2	311.2	575.8	832.9
E	Total electricity, gas and water supply		207	209	205	205		116.8	113.9	112.8	111.6		1927.8	2695.3	5083.4	7197.9

a/ Data were computed from the reported monthly average wages.
b/ 101 includes 102 and 103.
c/ 111 includes 112.
d/ 131 includes 132.
e/ 401 includes 402 and 403.

Belarus

ISIC Revision 3		Output (valuation not defined)					Value added (valuation not defined)				
			(billions of Belarusian Roubles)					(billions of Belarusian Roubles)			
ISIC	Industry	Note	2010	2011	2012	2013	Note	2010	2011	2012	2013
10	Mining of coal and lignite; extraction of peat		...	...	...	...		...	...	...	...
11	Extraction of crude oil and natural gas, etc.		...	...	...	...		...	...	...	...
12	Mining of uranium and thorium ores		...	...	...	...		...	...	...	...
13	Mining of metal ores		...	...	...	...		...	...	...	...
14	Other mining and quarrying		...	...	...	...		...	...	...	...
C	Total mining and quarrying		1397.5	5114.7	8678.1	9808.4		561.9	3372.3	5660.4	6372.5
40	Electricity, gas, steam and hot water supply		...	...	...	...		...	...	...	...
41	Water collection, purification, distribution		...	...	...	...		...	...	...	...
E	Total electricity, gas and water supply		16009.5	26334.8	43663.2	49169.4		4820.0	5081.1	15243.5	16689.4

Belgium

ISIC Revision 4		Number of enterprises					Number of employees					Wages and salaries paid to employees				
			(number)					(number)					(millions of Euros)			
ISIC	Industry	Note	2010	2011	2012	2013	Note	2010	2011	2012	2013	Note	2010	2011	2012	2013
05	Mining of coal and lignite		...	...	...	-		...	...	...	-		...	...	...	-
051	Mining of hard coal		...	...	-	-		...	...	-	-		...	...	-	-
052	Mining of lignite		-	-	...	-		-	-	...	-		-	-	...	-
06	Extraction of crude petroleum & natural gas		-	-	-	-		-	-	-	-		-	-	-	-
061	Extraction of crude petroleum		-	-	-	-		-	-	-	-		-	-	-	-
062	Extraction of natural gas		-	-	-	-		-	-	-	-		-	-	-	-
07	Mining of metal ores		...	...	-	-		...	...	-	-		...	...	-	-
071	Mining of iron ores		-	-	-	-		-	-	-	-		-	-	-	-
072	Mining of non-ferrous metal ores		...	...	-	-		...	...	-	-		...	...	-	-
08	Other mining and quarrying		252	227	...	177		2733	2874	...	2525		101	111	...	98
081	Quarrying of stone, sand and clay		234	227a/	166	169		2643	2874a/	2473	2350		98	111a/	100	92
089	Mining and quarrying n.e.c.		18	...a/	...	8		89	...a/	...	175		3	...a/	...	7
09	Mining support service activities		...	...	31	37		...	...	16	24		...	...	1	2
091	Support activities for petroleum		...	...	9	37b/		...	...	1	24b/		...	...	-	2b/
099	Support activities for other mining, quarrying		27	31	22	...b/		67	45	15	...b/		2	2	1	...b/
B	Mining and quarrying		293	274	203	214		2804	2922	2598	2549		103	114	105	100
35	Electricity, gas, steam & air conditioning		301	429	533	618		18819	17962	18007	19510		1297	1603	1445	1549
351	Electric power generation, transmission		...	...	501	577		...	...	16740	18076		...	...	1343	1437
352	Manufacture of gas		...	...	16	18		...	...	1251	1418		...	...	102	112
353	Steam and air conditioning supply		24	23	16	23		21	15	16	16		1	-	-	1
D	Electricity, gas, steam & air conditioning		301	429	533	618		18819	17962	18007	19510		1297	1603	1445	1549
36	Water collection, treatment and supply		...	61	40	31		...	8003	8184	7541		...	408	422	398
37	Sewerage		...	...	266	177		...	...	1863	2188		...	...	83	102
38	Waste collection,treatment,disposal activities		969	...	947	937		14979	...	14362	15217		551	...	539	587
381	Waste collection		151	...	174	186		4565	...	4157	4198		155	...	143	149
382	Waste treatment and disposal		281	390	318	276		6633	6070	6731	7353		262	237	284	313
383	Materials recovery		538	578	455	475		3782	3739	3474	3666		134	115	112	124
39	Remediation activities		139	75	76	95		613	839	1322	1298		26	35	59	58
E	Water supply;sewerage,waste management		1414	1519	1329	1240		23868	24401	25731	26244		1038	998	1103	1146

a/ 081 includes 089.
b/ 091 includes 099.

Belgium

ISIC Revision 4			Output (valuation not defined)					Value added at factor values			
			(millions of Euros)					(millions of Euros)			
ISIC	Industry	Note	2010	2011	2012	2013	Note	2010	2011	2012	2013
05	Mining of coal and lignite		...	...	...	-		...	...	...	-
051	Mining of hard coal		...	...	-	-		...	...	-	-
052	Mining of lignite		-	-	...	-		-	-	...	-
06	Extraction of crude petroleum & natural gas		-	-	-	-		-	-	-	-
061	Extraction of crude petroleum		-	-	-	-		-	-	-	-
062	Extraction of natural gas		-	-	-	-		-	-	-	-
07	Mining of metal ores		...	...	-	-		...	...	-	-
071	Mining of iron ores		-	-	-	-		-	-	-	-
072	Mining of non-ferrous metal ores		...	...	-	-		...	...	-	-
08	Other mining and quarrying		863	940	...	823		272	300	...	257
081	Quarrying of stone, sand and clay		825	940a/	779	716		266	300a/	256	242
089	Mining and quarrying n.e.c.		38	...a/	...	107		7	...a/	...	16
09	Mining support service activities		...	...	18	24		...	...	6	2
091	Support activities for petroleum		...	...	11	24b/		...	...	3	2b/
099	Support activities for other mining, quarrying		48	24	6	...b/		15	8	3	...b/
B	Mining and quarrying		938	977	857	847		290	311	271	259
35	Electricity, gas, steam & air conditioning		41669	42022	41252	42099		6694	7131	5677	7455
351	Electric power generation, transmission		...	...	35306	34830		...	...	5153	5315
352	Manufacture of gas		...	...	5889	7214		...	...	520	2135
353	Steam and air conditioning supply		83	49	57	55		2	4	5	5
D	Electricity, gas, steam & air conditioning		41669	42022	41252	42099		6694	7131	5677	7455
36	Water collection, treatment and supply		...	2578	2761	2661		...	927	1158	1088
37	Sewerage		...	...	730	1167		...	...	410	580
38	Waste collection,treatment,disposal activities		5205	...	5749	5804		1339	...	1373	1414
381	Waste collection		771	...	743	884		272	...	262	340
382	Waste treatment and disposal		1769	1968	2178	2179		662	655	779	751
383	Materials recovery		2666	3017	2828	2741		404	388	331	323
39	Remediation activities		208	338	425	393		40	35	60	61
E	Water supply;sewerage,waste management		8350	9155	9665	10025		2626	2556	3000	3143

a/ 081 includes 089.
b/ 091 includes 099.

Bosnia and Herzegovina

ISIC Revision 4		Number of enterprises					Number of employees					Wages and salaries paid to employees				
			(number)					(number)					(millions of Bosnian Convertible Marka)			
ISIC	Industry	Note	2010	2011	2012	2013	Note	2010	2011	2012	2013	Note	2010	2011	2012	2013
05	Mining of coal and lignite		...	19	18	...		...	14853	14988	...		...	178	...	...
051	Mining of hard coal		...	19a/	18a/	...		...	14853a/	14988a/	...		...	178a/	...	...
052	Mining of lignite		...	...a/	...a/	...		...	...a/	...a/	...		...	...a/	...	...
06	Extraction of crude petroleum & natural gas		...	...	...	...		...	...	...	...		...	...	...	...
061	Extraction of crude petroleum		...	...	...	...		...	...	...	...		...	...	...	...
062	Extraction of natural gas		...	...	...	...		...	...	...	...		...	...	...	...
07	Mining of metal ores		...	21	20	...		...	1849	2336	...		...	19	...	...
071	Mining of iron ores		...	21b/	20b/	...		...	1849b/	2336b/	...		...	19b/	...	...
072	Mining of non-ferrous metal ores		...	...b/	...b/	...		...	...b/	...b/	...		...	...b/	...	...
08	Other mining and quarrying		...	165	...	...		...	...	...	...		...	...	...	...
081	Quarrying of stone, sand and clay		...	165	140	...		...	1979	1745	...		...	13	...	...
089	Mining and quarrying n.e.c.		...	...	...	...		...	...	...	...		...	...	...	...
09	Mining support service activities		...	...	...	...		...	...	...	...		...	...	...	...
091	Support activities for petroleum		...	...	...	...		...	...	...	...		...	...	...	...
099	Support activities for other mining, quarrying		...	...	...	...		...	...	...	...		...	...	...	...
B	Mining and quarrying		...	223	194	...		...	19283	19884	...		...	216	...	...
35	Electricity, gas, steam & air conditioning		...	137	118	...		...	16988	17288	...		...	262	...	...
351	Electric power generation, transmission		...	104	83	...		...	15257	15428	...		...	241	...	...
352	Manufacture of gas		...	...	...	...		...	...	...	...		...	...	...	...
353	Steam and air conditioning supply		...	...	...	...		...	...	...	...		...	...	...	...
D	Electricity, gas, steam & air conditioning		...	137c/	118c/	...		...	16988c/	17288c/	...		...	262c/	...	...
36	Water collection, treatment and supply		...	144	142	...		...	6986	7091	...		...	65	...	...
37	Sewerage		...	11	11	...		...	232	222	...		...	2	...	...
38	Waste collection,treatment,disposal activities		...	178	178	...		...	5470	5571	...		...	44	...	...
381	Waste collection		...	72	72	...		...	4338	4495	...		...	35	...	...
382	Waste treatment and disposal		...	15	20	...		...	163	181	...		...	2	...	...
383	Materials recovery		...	91	86	...		...	969	896	...		...	7	...	...
39	Remediation activities		...	10	12	...		...	106	137	...		...	1	...	...
E	Water supply;sewerage,waste management		...	343c/	343c/	...		...	12794c/	13022c/	...		...	112c/	...	...

a/ 051 includes 052.
b/ 071 includes 072.
c/ Sum of available data.

Bosnia and Herzegovina

ISIC Revision 4		Output (valuation not defined)					Value added at factor values				
			(millions of Bosnian Convertible Marka)					(millions of Bosnian Convertible Marka)			
ISIC	Industry	Note	2010	2011	2012	2013	Note	2010	2011	2012	2013
05	Mining of coal and lignite		...	322	...	...		...	214	...	...
051	Mining of hard coal		...	322a/	...	...		...	214a/	...	...
052	Mining of lignite		...	...a/	...	...		...	...a/	...	...
06	Extraction of crude petroleum & natural gas		...	...	...	...		...	...	...	...
061	Extraction of crude petroleum		...	...	...	...		...	...	...	...
062	Extraction of natural gas		...	...	...	...		...	...	...	...
07	Mining of metal ores		...	86	...	...		...	30	...	...
071	Mining of iron ores		...	86b/	...	...		...	30b/	...	...
072	Mining of non-ferrous metal ores		...	...b/	...	...		...	...b/	...	...
08	Other mining and quarrying		...	...	...	...		...	...	...	...
081	Quarrying of stone, sand and clay		...	65	...	...		...	25	...	...
089	Mining and quarrying n.e.c.		...	...	...	...		...	...	...	...
09	Mining support service activities		...	...	...	...		...	...	...	...
091	Support activities for petroleum		...	...	...	...		...	...	...	...
099	Support activities for other mining, quarrying		...	...	...	...		...	...	...	...
B	Mining and quarrying		...	493	...	...		...	279	...	...
35	Electricity, gas, steam & air conditioning		...	1051	...	...		...	517	...	...
351	Electric power generation, transmission		...	906	...	...		...	495	...	...
352	Manufacture of gas		...	...	...	...		...	...	...	...
353	Steam and air conditioning supply		...	...	...	...		...	...	...	...
D	Electricity, gas, steam & air conditioning		...	1051c/	...	...		...	517c/	...	...
36	Water collection, treatment and supply		...	119	...	...		...	80	...	...
37	Sewerage		...	3	...	...		...	2	...	...
38	Waste collection,treatment,disposal activities		...	224	...	...		...	69	...	...
381	Waste collection		...	65	...	...		...	45	...	...
382	Waste treatment and disposal		...	4	...	...		...	3	...	...
383	Materials recovery		...	154	...	...		...	22	...	...
39	Remediation activities		...	3	...	...		...	2	...	...
E	Water supply;sewerage,waste management		...	350c/	...	...		...	153c/	...	...

a/ 051 includes 052.
b/ 071 includes 072.
c/ Sum of available data.

Brazil

ISIC Revision 4		Number of establishments					Number of employees					Wages and salaries paid to employees			
			(number)					(number)					(millions of Brazilian Reais)		
ISIC	Industry	Note	2009	2010	2011	2012	Note	2009	2010	2011	2012	Note	2009	2010	2011	2012
051	Mining of hard coal		...	...	...	...		...	...	...	...		...	...	...	...
052	Mining of lignite		...	...	...	...		...	...	...	...		...	...	...	...
061	Extraction of crude petroleum		...	...	...	...		...	...	...	...		...	...	...	...
062	Extraction of natural gas		...	...	...	...		...	...	...	...		...	...	...	...
071	Mining of iron ores		169	153	160	174		37518	41471	52925	60683		1627.6	2633.6	2942.5	3637.9
072	Mining of non-ferrous metal ores		193	189	205	254		14719	17647	21178	23181		516.7	732.4	835.4	1016.9
081	Quarrying of stone, sand and clay		3159	3146	3205	3129		57687	59718	66194	65979		842.9	960.5	1239.9	1353.5
089	Mining and quarrying n.e.c.		463	430	457	476		17443	18360	18454	18352		310.0	418.2	442.0	471.8
091	Support activities for petroleum		176	201	233	298		26050	28164	33259	37542		1701.3	2002.8	2675.6	3391.5
099	Support activities for other mining, quarrying		3	23	29	22		26	243	1172	1313		0.3	3.4	24.5	35.8
B	Mining and quarrying	a/	4163	4142	4289	4353	a/	153443	165603	193182	207050	a/	4998.7	6750.9	8160.0	9907.4
351	Electric power generation, transmission		...	...	...	...		...	...	...	...		...	...	...	...
352	Manufacture of gas		...	...	...	...		...	...	...	...		...	...	...	...
353	Steam and air conditioning supply		...	...	...	...		...	...	...	...		...	...	...	...
D	Electricity, gas, steam & air conditioning		...	...	...	...		...	...	...	...		...	...	...	...
360	Water collection, treatment and supply		...	...	...	...		...	...	...	...		...	...	...	...
370	Sewerage		...	...	...	...		...	...	...	...		...	...	...	...
381	Waste collection		...	...	...	...		...	...	...	...		...	...	...	...
382	Waste treatment and disposal		...	...	...	...		...	...	...	...		...	...	...	...
383	Materials recovery		...	...	...	...		...	...	...	...		...	...	...	...
390	Remediation activities		...	...	...	...		...	...	...	...		...	...	...	...
E	Water supply;sewerage,waste management		...	...	...	...		...	...	...	...		...	...	...	...

a/ Sum of available data.

Brazil

ISIC Revision 4		Output at factor values					Value added at factor values				
			(millions of Brazilian Reais)					(millions of Brazilian Reais)			
ISIC	Industry	Note	2009	2010	2011	2012	Note	2009	2010	2011	2012
051	Mining of hard coal		...	...	...	...		...	...	...	...
052	Mining of lignite		...	...	...	...		...	...	...	...
061	Extraction of crude petroleum		...	...	...	...		...	...	...	...
062	Extraction of natural gas		...	...	...	...		...	...	...	...
071	Mining of iron ores		28744.3	60852.3	80166.7	70866.6		20656.1	49521.3	66561.6	60255.7
072	Mining of non-ferrous metal ores		5173.4	7257.5	8278.4	9451.0		2534.4	3272.5	4179.5	4812.4
081	Quarrying of stone, sand and clay		5773.1	7181.2	9062.3	9947.0		3421.8	4366.5	5389.4	6070.5
089	Mining and quarrying n.e.c.		3504.7	4158.1	4259.4	4366.3		1898.1	2216.2	2201.3	2774.0
091	Support activities for petroleum		6755.0	7652.2	9409.4	11917.2		4843.8	5676.8	7211.4	8927.6
099	Support activities for other mining, quarrying		0.6	13.2	146.5	180.1		0.6	9.1	127.7	149.4
B	Mining and quarrying	a/	49951.1	87114.5	111322.8	106728.2	a/	33354.8	65062.5	85670.9	82989.5
351	Electric power generation, transmission		...	...	...	...		...	...	...	...
352	Manufacture of gas		...	...	...	...		...	...	...	...
353	Steam and air conditioning supply		...	...	...	...		...	...	...	...
D	Electricity, gas, steam & air conditioning		...	...	...	...		...	...	...	...
360	Water collection, treatment and supply		...	...	...	...		...	...	...	...
370	Sewerage		...	...	...	...		...	...	...	...
381	Waste collection		...	...	...	...		...	...	...	...
382	Waste treatment and disposal		...	...	...	...		...	...	...	...
383	Materials recovery		...	...	...	...		...	...	...	...
390	Remediation activities		...	...	...	...		...	...	...	...
E	Water supply;sewerage,waste management		...	...	...	...		...	...	...	...

a/ Sum of available data.

Bulgaria

ISIC Revision 4		Number of establishments					Number of employees					Wages and salaries paid to employees				
			(number)					(number)					(thousands of Bulgarian Leva)			
ISIC	Industry	Note	2010	2011	2012	2013	Note	2010	2011	2012	2013	Note	2010	2011	2012	2013
051	Mining of hard coal		3	3	3	4		...	...	153	128		...	...	1170	1083
052	Mining of lignite		20	22	21	20		13062	13093	12766	12375		166593	187393	197008	187304
061	Extraction of crude petroleum		3	3	...	...		...	...	...	...		...	...	...	...
062	Extraction of natural gas		4	6	...	...		33	48	46	57		1024	1276	1510	...
071	Mining of iron ores		-	-	...	...		-	-	...	...		-	-	...	...
072	Mining of non-ferrous metal ores		21	22	24	26		6332	6573	6718	6721		94363	109093	120705	124944
081	Quarrying of stone, sand and clay		261	254	259	266		3831	3775	3736	3700		28715	30754	31772	31742
089	Mining and quarrying n.e.c.		41	35	32	30		403	330	478	464		3357	3195	4739	4910
091	Support activities for petroleum		6	10	12	11		9	19	25	35		402	705	902	1366
099	Support activities for other mining, quarrying		19	24	24	23		135	147	159	182		1276	1699	1906	2181
B	Mining and quarrying		378	379	382	386		24584	24743	24641	24268		301844	340191	364644	361274
351	Electric power generation, transmission		1008	1319	1631	1714		27325	26677	27464	27148		476852	492897	531945	524369
352	Manufacture of gas		43	40	39	38		1035	778	755	713		15028	10727	12695	11500
353	Steam and air conditioning supply		40	36	33	32		4068	3723	3593	3476		45032	42198	41961	42345
D	Electricity, gas, steam & air conditioning		1091	1395	1703	1784		32428	31178	31812	31337		536912	545822	586601	578214
360	Water collection, treatment and supply		81	78	71	70		18023	18100	17183	17137		148752	151998	145488	151429
370	Sewerage		63	59	67	64		253	224	555	560		1909	1739	3709	4246
381	Waste collection		328	336	328	320		12248	11933	11932	12031		79554	79292	80773	83243
382	Waste treatment and disposal		53	67	80	75		1536	2046	2168	2535		14431	21802	23456	34770
383	Materials recovery		127	144	149	161		2277	2569	2516	2241		15880	20803	22825	19235
390	Remediation activities		50	51	53	61		412	687	711	718		3463	6053	6230	7143
E	Water supply;sewerage,waste management		702	735	748	751		34749	35559	35065	35222		263989	281687	282481	300066

Bulgaria

ISIC Revision 4		Output at producers' prices					Value added at factor values				
		Note	(millions of Bulgarian Leva)				Note	(millions of Bulgarian Leva)			
ISIC	Industry		2010	2011	2012	2013		2010	2011	2012	2013
051	Mining of hard coal		5	9	6	6		2	1	2	...
052	Mining of lignite		647	799	709	587		373	470	358	...
061	Extraction of crude petroleum		...	...	...	...		...	...	...	...
062	Extraction of natural gas		...	...	...	...		...	...	...	...
071	Mining of iron ores		-	-	...	...		-	-	...	...
072	Mining of non-ferrous metal ores		1324	1513	1743	1505		731	867	1021	767
081	Quarrying of stone, sand and clay		224	236	232	230		83	79	83	78
089	Mining and quarrying n.e.c.		19	21	28	36		4	7	8	9
091	Support activities for petroleum		...	12	4	3		...	...	1	2
099	Support activities for other mining, quarrying		...	11	15	16		...	...	4	5
B	Mining and quarrying		2286	2802	2970	2557		1205	1584	1660	1299
351	Electric power generation, transmission		6134	7405	8475	7808		2413	2608	2772	2979
352	Manufacture of gas		72	46	4	167		19	5	-37	112
353	Steam and air conditioning supply		888	1081	1167	1068		184	250	225	232
D	Electricity, gas, steam & air conditioning		7094	8532	9646	9043		2616	2863	2960	3323
360	Water collection, treatment and supply		621	632	639	643		322	338	336	335
370	Sewerage		9	11	12	13		4	5	5	5
381	Waste collection		320	346	351	333		150	154	159	161
382	Waste treatment and disposal		66	73	86	92		42	58	56	80
383	Materials recovery		345	237	226	184		97	94	95	77
390	Remediation activities		29	59	60	93		13	16	15	17
E	Water supply;sewerage,waste management		1390	1358	1374	1358		628	665	666	675

Burundi

ISIC Revision 3		Number of enterprises					Number of employees					Wages and salaries paid to employees			
			(number)					(number)					(millions of Burundi Francs)		
ISIC	Industry	Note	2009	2010	2011	2012	Note	2009	2010	2011	2012	Note	2009	2010	2011	2012
101	Mining and agglomeration of hard coal		...	...	...	...		...	...	...	...		...	...	...	...
102	Mining and agglomeration of lignite		...	...	...	...		...	...	...	...		...	...	...	...
103	Extraction and agglomeration of peat		...	...	...	...		...	...	...	...		...	...	...	...
111	Extraction of crude oil and natural gas		...	...	...	...		...	...	...	...		...	...	...	...
112	Oil, gas service activities		...	...	...	...		...	...	...	...		...	...	...	...
120	Mining of uranium and thorium ores		...	...	...	...		...	...	...	...		...	...	...	...
131	Iron ores		...	...	...	...		...	...	...	...		...	...	...	...
132	Non-ferrous metal ores,excl.uranium,thorium		...	...	...	...		...	...	...	...		...	...	...	...
141	Quarrying of stone, sand and clay		...	...	...	...		...	...	...	...		...	...	...	...
142	Mining and quarrying n.e.c.		...	...	...	...		...	...	...	...		...	...	...	...
C	Total mining and quarrying		...	...	...	...		2216	...	...	...		254	360	...	...
401	Electricity production, collection, distribution		...	...	1	1		...	...	1015	1038		4201	4833	6171	6310
402	Gas production and distribution		...	...	...	...		...	...	...	...		...	...	...	...
403	Steam and hot water supply		...	...	...	...		...	...	...	...		...	...	...	...
410	Water collection, purification, distribution		...	...	1	1		...	...	435	445		2801	3222	2270	3559
E	Total electricity, gas and water supply		...	...	2a/	2a/		1391	1530	1450a/	1483a/	a/	7002	8055	8441	9869

a/ Sum of available data.

Burundi

ISIC Revision 3		Output (valuation not defined) a/					Value added (valuation not defined)				
			(millions of Burundi Francs)					(millions of Burundi Francs)			
ISIC	Industry	Note	2009	2010	2011	2012	Note	2009	2010	2011	2012
101	Mining and agglomeration of hard coal		...	...	...	...		...	...	...	...
102	Mining and agglomeration of lignite		...	...	...	...		...	...	...	...
103	Extraction and agglomeration of peat		...	...	...	...		...	...	...	...
111	Extraction of crude oil and natural gas		...	...	...	...		...	...	...	...
112	Oil, gas service activities		...	...	...	...		...	...	...	...
120	Mining of uranium and thorium ores		...	...	...	...		...	...	...	...
131	Iron ores		...	...	...	...		...	...	...	...
132	Non-ferrous metal ores,excl.uranium,thorium		...	...	...	...		...	...	...	...
141	Quarrying of stone, sand and clay		...	...	...	...		...	...	...	...
142	Mining and quarrying n.e.c.		...	...	...	...		...	...	...	...
C	Total mining and quarrying		12738	12548	...	...		12334	12153	...	...
401	Electricity production, collection, distribution		12722	13729	17940	26093		9632	6921	8565	15057
402	Gas production and distribution		...	...	...	...		...	...	...	...
403	Steam and hot water supply		...	...	...	...		...	...	...	...
410	Water collection, purification, distribution		8481	9153	11960	17362		6421	4841	6133	10038
E	Total electricity, gas and water supply	b/	21203	22882	29900	43455	b/	16053	11762	14698	25095

a/ Output is reported at basic prices.
b/ Sum of available data.

Canada

ISIC Revision 4		Number of establishments					Number of employees					Wages and salaries paid to employees				
			(number)					(number)					(thousands of Canadian Dollars)			
ISIC	Industry	Note	2010	2011	2012	2013	Note	2010	2011	2012	2013	Note	2010	2011	2012	2013
051	Mining of hard coal		...	...	...	...		...	...	...	...		...	...	...	...
052	Mining of lignite		...	...	...	13		...	...	...	...		...	...	...	...
061	Extraction of crude petroleum		...	...	...	5608a/		...	54113a/	58887a/	61572a/		...	...	...	...
062	Extraction of natural gas		...	...	...	...a/		...	...a/	...a/	...a/		...	...	...	...
071	Mining of iron ores		...	...	...	15		...	...	...	...		...	...	...	...
072	Mining of non-ferrous metal ores		...	...	...	...		...	...	...	...		...	...	...	...
081	Quarrying of stone, sand and clay		...	...	...	1252b/		...	...	...	...		...	...	...	...
089	Mining and quarrying n.e.c.		161	152	152	...b/		3370	...	...	...		161218	156656	169195	...
091	Support activities for petroleum		...	...	...	8709		...	96250c/	98906c/	99435c/		...	...	...	...
099	Support activities for other mining, quarrying		...	...	...	4278		...	...c/	...c/	...c/		...	...	...	...
B	Mining and quarrying		...	...	...	22330		...	207051	221232	226020		...	...	...	...
351	Electric power generation, transmission		...	...	...	1658		...	...	...	...		...	...	...	...
352	Manufacture of gas		...	...	...	438		...	...	...	...		...	...	...	...
353	Steam and air conditioning supply		...	...	...	63		...	...	...	...		...	...	...	...
D	Electricity, gas, steam & air conditioning		...	...	...	...		...	123465d/	122398d/	118889d/		...	...	...	...
360	Water collection, treatment and supply		...	...	...	...		...	...	...	...		...	...	...	...
370	Sewerage		...	...	...	83		...	...	...	...		...	...	...	...
381	Waste collection		...	...	...	...		...	...	...	...		...	...	...	...
382	Waste treatment and disposal		...	...	...	...		...	...	...	...		...	...	...	...
383	Materials recovery		...	...	...	...		...	...	...	...		...	...	...	...
390	Remediation activities		...	...	...	...		...	...	...	...		...	...	...	...
E	Water supply;sewerage,waste management		...	...	...	857		...	...d/	...d/	...d/		...	...	...	...

a/ 061 includes 062.
b/ 081 includes 089.
c/ 091 includes 099.
d/ D includes E.

China

ISIC Revision 3		Number of enterprises					Employees					Wages and salaries				
			(number)					(thousands)					(millions of Chinese Yuan)			
ISIC	Industry	Note	2010	2011	2012	2013	Note	2010	2011	2012	2013	Note	2010	2011	2012	2013
101	Mining and agglomeration of hard coal		8725	7455	...	7929a/		...	...	...	...		...	...	...	...
102	Mining and agglomeration of lignite		228	204	...	...a/		...	...	...	...		...	...	...	...
103	Extraction and agglomeration of peat		200	134	...	...a/		...	...	...	...		...	...	...	...
111	Extraction of crude oil and natural gas		143	137	...	144b/		...	...	...	...		...	...	...	...
112	Oil, gas service activities		167	134	...	...b/		...	...	...	...		...	...	...	...
120	Mining of uranium and thorium ores		8	7	...	...		...	...	...	...		...	...	...	...
131	Iron ores		3930	3248	...	3614		...	...	...	...		...	...	...	...
132	Non-ferrous metal ores,excl.uranium,thorium		2646	2218	...	2108		...	...	...	...		...	...	...	...
141	Quarrying of stone, sand and clay		3429	2342	...	3717		...	...	...	...		...	...	...	...
142	Mining and quarrying n.e.c.		1364	1024	...	199		...	...	...	...		...	...	...	...
C	Total mining and quarrying	c/	20840	16903	...	17711		...	...	...	...		...	...	...	...
401	Electricity production, collection, distribution		5504	4483	...	6010d/		...	...	...	...		...	...	...	...
402	Gas production and distribution		970	875	...	1132		...	...	...	...		...	...	...	...
403	Steam and hot water supply		1054	804	...	...d/		...	...	...	...		...	...	...	...
410	Water collection, purification, distribution		2109	1153	...	1376		...	...	...	...		...	...	...	...
E	Total electricity, gas and water supply	c/	9637	7315	...	8518		...	...	...	...		...	...	...	...

a/ 101 includes 102 and 103.
b/ 111 includes 112.
c/ Sum of available data.
d/ 401 includes 403.

China

ISIC Revision 3		Output at producers' prices					Value added				
			(billions of Chinese Yuan)					(billions of Chinese Yuan)			
ISIC	Industry	Note	2010	2011	2012	2013	Note	2010	2011	2012	2013
101	Mining and agglomeration of hard coal		2194.4	2898.3	...	...		...	...	...	...
102	Mining and agglomeration of lignite		162.7	238.4	...	...		...	...	...	...
103	Extraction and agglomeration of peat		12.3	20.0	...	...		...	...	...	...
111	Extraction of crude oil and natural gas		893.4	1097.5	...	...		...	...	...	...
112	Oil, gas service activities		168.3	190.7	...	...		...	...	...	...
120	Mining of uranium and thorium ores		1.0	1.0	...	...		...	...	...	...
131	Iron ores		583.2	773.2	...	...		...	...	...	...
132	Non-ferrous metal ores,excl.uranium,thorium		400.0	517.7	...	...		...	...	...	...
141	Quarrying of stone, sand and clay		200.7	251.2	...	...		...	...	...	...
142	Mining and quarrying n.e.c.		115.7	137.0	...	...		...	...	...	...
C	Total mining and quarrying	a/	4731.8	6125.0	...	...		...	...	...	...
401	Electricity production, collection, distribution		3964.1	4608.4	...	...		...	...	...	...
402	Gas production and distribution		250.6	320.5	...	...		...	...	...	...
403	Steam and hot water supply		92.0	101.3	...	...		...	...	...	...
410	Water collection, purification, distribution		114.3	116.7	...	...		...	...	...	...
E	Total electricity, gas and water supply	a/	4421.0	5147.0	...	...		...	...	...	...

a/ Sum of available data.

China, Hong Kong SAR

ISIC Revision 4		Number of establishments					Number of employees					Wages and salaries paid to employees				
			(number)					(number)					(millions of Hong Kong Dollars)			
ISIC	Industry	Note	2010	2011	2012	2013	Note	2010	2011	2012	2013	Note	2010	2011	2012	2013
05	Mining of coal and lignite		...	...	...	...		...	...	...	...		...	...	...	...
06	Extraction of crude petroleum & natural gas		...	...	...	...		...	...	...	...		...	...	...	...
07	Mining of metal ores		...	...	...	...		...	...	...	...		...	...	...	...
08	Other mining and quarrying		2	2	2	2		100	100	100	100		...	...	...	...
09	Mining support service activities		...	...	...	...		...	...	...	...		...	...	...	...
B	Mining and quarrying		2	2	2	2		100	100	100	100		...	...	...	...
35	Electricity, gas, steam & air conditioning		25	19	12	8		7700	7700	7700	7700		3641	3771	3975	4235
D	Electricity, gas, steam & air conditioning		25	19	12	8		7700	7700	7700	7700		3641	3771	3975	4235
36	Water collection, treatment and supply		...	...	...	...		...	...	...	...		...	...	...	...
37	Sewerage		...	...	...	...		...	...	...	...		...	...	...	...
38	Waste collection,treatment,disposal activities		...	...	...	...		...	...	...	...		...	...	...	...
39	Remediation activities		...	...	...	...		...	...	...	...		...	...	...	...
E	Water supply;sewerage,waste management	a/	300	296	302	321	a/	3300	3200	3100	3200	a/	583	591	604	645

a/ E excludes 36.

China, Hong Kong SAR

ISIC Revision 4		Output at producers' prices					Value added at producers' prices				
			(millions of Hong Kong Dollars)					(millions of Hong Kong Dollars)			
ISIC	Industry	Note	2010	2011	2012	2013	Note	2010	2011	2012	2013
05	Mining of coal and lignite		...	...	...	...		...	...	...	...
06	Extraction of crude petroleum & natural gas		...	...	...	...		...	...	...	...
07	Mining of metal ores		...	...	...	...		...	...	...	...
08	Other mining and quarrying		...	...	...	...		...	...	...	...
09	Mining support service activities		...	...	...	...		...	...	...	...
B	Mining and quarrying		...	...	...	...		...	...	...	...
35	Electricity, gas, steam & air conditioning		50523	52782	54696	54196		32101	31676	33352	32425
D	Electricity, gas, steam & air conditioning		50523	52782	54696	54196		32101	31676	33352	32425
36	Water collection, treatment and supply		...	...	...	...		...	...	...	...
37	Sewerage		...	...	...	...		...	...	...	...
38	Waste collection,treatment,disposal activities		...	...	...	...		...	...	...	...
39	Remediation activities		...	...	...	...		...	...	...	...
E	Water supply;sewerage,waste management	a/	2163	2568	2889	3061	a/	977	1133	1175	1238

a/ E excludes 36.

China, Macao SAR

ISIC Revision 3		Number of establishments					Number of persons engaged					Wages and salaries paid to employees			
		Note	(number)				Note	(number)				Note	(thousands of Macao Patacas)			
ISIC	Industry		2010	2011	2012	2013		2010	2011 a/	2012 a/	2013 a/		2010	2011	2012	2013
101	Mining and agglomeration of hard coal		...	...	...	...		...	...	...	...		...	...	...	...
102	Mining and agglomeration of lignite		...	...	...	...		...	...	...	...		...	...	...	...
103	Extraction and agglomeration of peat		...	...	...	...		...	...	...	...		...	...	...	...
111	Extraction of crude oil and natural gas		...	...	...	...		...	...	...	...		...	...	...	...
112	Oil, gas service activities		...	...	...	...		...	...	...	...		...	...	...	...
120	Mining of uranium and thorium ores		...	...	...	...		...	...	...	...		...	...	...	...
131	Iron ores		...	...	...	...		...	...	...	...		...	...	...	...
132	Non-ferrous metal ores,excl.uranium,thorium		...	...	...	...		...	...	...	...		...	...	...	...
141	Quarrying of stone, sand and clay		1	1	1	1		17	19	23	9		3489	4799	7105	3101
142	Mining and quarrying n.e.c.		...	...	...	...		...	...	...	...		...	...	...	...
C	Total mining and quarrying		1	1	1	1		17	19	23	9		3489	4799	7105	3101
401	Electricity production, collection, distribution		1	1	1	1		725	728	...	725		279169	289523	...	459867
402	Gas production and distribution		1	1	-	2		...	...	-	...		...	...	-	...
403	Steam and hot water supply		2	2	2	2		...	...	...	...		...	...	...	...
410	Water collection, purification, distribution		1	1	1	1		257	257	263	259		64044	67070	70110	88029
E	Total electricity, gas and water supply		5	5	4	6		1064	1065	1020	1097		354703	368568	373511	567486

a/ Number of employees.

China, Macao SAR

ISIC Revision 3		Output at producers' prices					Value added at producers' prices				
			(millions of Macao Patacas)					(millions of Macao Patacas)			
ISIC	Industry	Note	2010	2011	2012	2013	Note	2010	2011	2012	2013
101	Mining and agglomeration of hard coal		...	...	...	...		...	...	...	...
102	Mining and agglomeration of lignite		...	...	...	...		...	...	...	...
103	Extraction and agglomeration of peat		...	...	...	...		...	...	...	...
111	Extraction of crude oil and natural gas		...	...	...	...		...	...	...	...
112	Oil, gas service activities		...	...	...	...		...	...	...	...
120	Mining of uranium and thorium ores		...	...	...	...		...	...	...	...
131	Iron ores		...	...	...	...		...	...	...	...
132	Non-ferrous metal ores,excl.uranium,thorium		...	...	...	...		...	...	...	...
141	Quarrying of stone, sand and clay		9.7	27.7	35.0	4.3		4.0	13.3	17.2	3.0
142	Mining and quarrying n.e.c.		...	...	...	...		...	...	...	...
C	Total mining and quarrying		9.7	27.7	35.0	4.3		4.0	13.3	17.2	3.0
401	Electricity production, collection, distribution		2317.0	2308.8	...	2178.0		1425.0	1448.0	...	1737.9
402	Gas production and distribution		...	...	-	...		...	...	-	...
403	Steam and hot water supply		...	...	...	...		...	...	...	...
410	Water collection, purification, distribution		348.4	401.3	444.8	466.6		158.7	197.8	199.2	222.8
E	Total electricity, gas and water supply		3095.0	2919.2	2693.3	2663.1		1583.8	1629.6a/	1820.3	1962.7

a/ Includes groups with negative value added.

China, Taiwan Province

ISIC Revision 4		Output (valuation not defined)					Value added (valuation not defined)				
			(millions of Taiwan Dollars)					(millions of Taiwan Dollars)			
ISIC	Industry	Note	2010	2011	2012	2013	Note	2010	2011	2012	2013
051	Mining of hard coal		...	...	...	...		...	...	...	...
052	Mining of lignite		...	...	...	...		...	...	...	...
061	Extraction of crude petroleum		...	...	...	...		...	...	...	...
062	Extraction of natural gas		...	...	...	...		...	...	...	...
071	Mining of iron ores		...	...	...	...		...	...	...	...
072	Mining of non-ferrous metal ores		...	...	...	...		...	...	...	...
081	Quarrying of stone, sand and clay		...	...	...	...		...	...	...	...
089	Mining and quarrying n.e.c.		...	...	...	...		...	...	...	...
091	Support activities for petroleum		...	...	...	...		...	...	...	...
099	Support activities for other mining, quarrying		...	...	...	...		...	...	...	...
B	Mining and quarrying		31023	29489	28510	28711		18415	17174	15500	15597
351	Electric power generation, transmission		546150	561435	590591	623551		155382	127281	123251	174120
352	Manufacture of gas		96162	82485	105474	123875		27520	-1268	11644	29562
353	Steam and air conditioning supply		...	...	...	...		...	...	...	...
D	Electricity, gas, steam & air conditioning		642312	643920	696065	747426		182902	126013	134895	203682
360	Water collection, treatment and supply		42875	43188	43163	45016		22077	22493	22489	22475
370	Sewerage	a/	160186	166151	168826	169922	a/	73708	74450	78448	77142
381	Waste collection	a/	...	...	...	...	a/	...	...	...	...
382	Waste treatment and disposal	a/	...	...	...	...	a/	...	...	...	...
383	Materials recovery	a/	...	...	...	...	a/	...	...	...	...
390	Remediation activities	a/	...	...	...	...	a/	...	...	...	...
E	Water supply;sewerage,waste management		203061	209339	211989	214938		95785	96943	100937	99617

a/ 370 includes 381, 382, 383 and 390.

Croatia

ISIC Revision 4		Number of establishments					Number of employees					Wages and salaries paid to employees				
			(number)					(number)					(thousands of Croatian Kunas)			
ISIC	Industry	Note	2010	2011	2012	2013	Note	2010	2011	2012	2013	Note	2010	2011	2012	2013
051	Mining of hard coal		1	1	1	1		...	...	...	...		...	...	...	...
052	Mining of lignite		3	3	3	3		...	...	...	...		...	...	...	...
061	Extraction of crude petroleum		14	15	17	18		224	173	151	140	a/	275082	231776	208344	203748
062	Extraction of natural gas		4	5	5	5		1682	1383	1219	1184	a/	...	...	...	...
071	Mining of iron ores		-	-	-	-		-	-	-	-		-	-	-	-
072	Mining of non-ferrous metal ores		3	3	3	3		...	...	...	...		...	...	...	...
081	Quarrying of stone, sand and clay		401	400	362	367		2329	1871	1764	1707	b/	195622	159109	155564	155428
089	Mining and quarrying n.e.c.		48	46	45	42		237	203	199	180	b/	...	...	...	...
091	Support activities for petroleum		15	14	13	14		2889	2637	2257	2049	c/	378864	350159	207712	213264
099	Support activities for other mining, quarrying		4	4	4	4		3	3	3	2	c/	...	...	...	...
B	Mining and quarrying		493	491	453	457		7367	6273	5593	5262		880239	747415	1964616	1834543
351	Electric power generation, transmission		438	592	893	1159		13744	13597	13413	12547	d/	1972758	2008409	2030646	1987390
352	Manufacture of gas		91	88	91	94		2015	2049	2074	2066	d/	...	...	...	...
353	Steam and air conditioning supply		43	40	40	41		909	1054	1087	1054	d/	...	...	...	...
D	Electricity, gas, steam & air conditioning		572	720	1024	1294		16668	16700	16574	15667		1972758	2008409	2030646	1987390
360	Water collection, treatment and supply		189	193	192	204		9315	9394	9298	8962		832873	850645	850432	845511
370	Sewerage		33	36	36	45		302	239	342	334		...	...	22318	23363
381	Waste collection		331	352	354	406		8351	8520	8664	8484	e/	906243	918394	924469	938139
382	Waste treatment and disposal		53	61	73	108		158	183	218	291	e/	...	...	...	...
383	Materials recovery		395	397	328	344		2040	1862	1806	1730	e/	...	...	...	...
390	Remediation activities		78	86	85	97		1613	1681	1572	1660		146718	140760	134029	146970
E	Water supply;sewerage,waste management		1079	1125	1068	1204		21779	21879	21900	21461		1909931	1928415	1936836	1959561

a/ 061 includes 062.
b/ 081 includes 089.
c/ 091 includes 099.
d/ 351 includes 352 and 353.
e/ 381 includes 382 and 383.

Cyprus

ISIC Revision 4		Number of enterprises					Number of employees					Wages and salaries paid to employees			
			(number)					(number)					(thousands of Euros)		
ISIC	Industry	Note	2010	2011	2012	2013	Note	2010	2011	2012	2013	Note	2010	2011	2012	2013
051	Mining of hard coal		-	-	-	-		-	-	-	-		-	-	-	-
052	Mining of lignite		-	-	-	-		-	-	-	-		-	-	-	-
061	Extraction of crude petroleum		-	-	-	-		-	-	-	-		-	-	-	-
062	Extraction of natural gas		-	-	-	-		-	-	-	-		-	-	-	-
071	Mining of iron ores		-	-	-	-		-	-	-	-		-	-	-	-
072	Mining of non-ferrous metal ores		-	-	-	-		-	-	-	-		-	-	-	-
081	Quarrying of stone, sand and clay		90a/	...	...	...		546a/	...	...	...		14799a/	...	...	...
089	Mining and quarrying n.e.c.		...a/	...	...	...		...a/	...	...	...		...a/	...	...	...
091	Support activities for petroleum		-	...	...	...		-	...	...	...		-	...	...	...
099	Support activities for other mining, quarrying		-	-	-	-		-	-	-	-		-	-	-	-
B	Mining and quarrying		90	99b/	92b/	108b/		546	2999b/	2882b/	2744b/		14799	98070b/	97152b/	91804b/
351	Electric power generation, transmission		4	...	...	...		2470	...	...	...		83397	...	...	...
352	Manufacture of gas		-	-	-	-		-	-	-	-		-	-	-	-
353	Steam and air conditioning supply		-	-	-	-		-	-	-	-		-	-	-	-
D	Electricity, gas, steam & air conditioning		4	...b/	...b/	...b/		2470	...b/	...b/	...b/		83397	...b/	...b/	...b/
360	Water collection, treatment and supply		8	8	9	7		404	405	392	361		12072	12194	11514	10008
370	Sewerage		66	65	66	66		396	400	409	405		11442	12253	12202	10917
381	Waste collection		65	67	74	73		242	357	370	377		4489	6863	6782	6179
382	Waste treatment and disposal		11c/	11c/	11	12		98c/	84c/	71	101		1752c/	1433c/	1323	1503
383	Materials recovery		38	35	38	44d/		311	377	379	391d/		4862	5955	6076	6159d/
390	Remediation activities		...c/	...c/	-	...d/		...c/	...c/	-	...d/		...c/	...c/	-	...d/
E	Water supply;sewerage,waste management		188	186	198	202		1451	1623	1621	1635		34617	38698	37897	34766

a/ 081 includes 089.
b/ B includes D.
c/ 382 includes 390.
d/ 383 includes 390.

Cyprus

ISIC Revision 4			Output at producers' prices					Value added at producers' prices			
			(thousands of Euros)					(thousands of Euros)			
ISIC	Industry	Note	2010	2011	2012	2013	Note	2010	2011	2012	2013
051	Mining of hard coal		-	-	-	-		-	-	-	-
052	Mining of lignite		-	-	-	-		-	-	-	-
061	Extraction of crude petroleum		-	-	-	-		-	-	-	-
062	Extraction of natural gas		-	-	-	-		-	-	-	-
071	Mining of iron ores		-	-	-	-		-	-	-	-
072	Mining of non-ferrous metal ores		-	-	-	-		-	-	-	-
081	Quarrying of stone, sand and clay		86055a/	...	...	...		50255a/	...	...	...
089	Mining and quarrying n.e.c.		...a/	...	...	...		...a/	...	...	...
091	Support activities for petroleum		-	...	...	...		-	...	...	...
099	Support activities for other mining, quarrying		-	-	-	-		-	-	-	-
B	Mining and quarrying		86055	1022872b/	1197075b/	990095b/		50255	350130b/	398283b/	358036b/
351	Electric power generation, transmission		811948	...	...	...		328906	...	...	...
352	Manufacture of gas		-	-	-	-		-	-	-	-
353	Steam and air conditioning supply		-	-	-	-		-	-	-	-
D	Electricity, gas, steam & air conditioning		811948	...b/	...b/	...b/		328906	...b/	...b/	...b/
360	Water collection, treatment and supply		103361	106416	85283	80519		36440	38155	37385	38717
370	Sewerage		82743	88589	103002	104184		50863	48437	66314	74319
381	Waste collection		13168	26327	24336	19882		7593	15045	13980	11517
382	Waste treatment and disposal		8366c/	10659c/	9530	9401		5296c/	6167c/	4966	4731
383	Materials recovery		69179	102422	102268	76967d/		19234	20244	19673	16541d/
390	Remediation activities		...c/	...c/	-	...d/		...c/	...c/	-	...d/
E	Water supply;sewerage,waste management		276817	334413	324419	290953		119426	128048	142318	145825

a/ 081 includes 089.
b/ B includes D.
c/ 382 includes 390.
d/ 383 includes 390.

Czech Republic

ISIC Revision 4		Number of enterprises					Number of employees					Wages and salaries paid to employees				
			(number)					(thousands)					(millions of Czech Koruny)			
ISIC	Industry	Note	2010	2011	2012	2013	Note	2010	2011	2012	2013	Note	2010	2011	2012	2013
051	Mining of hard coal		5	5	5	6	a/	24.0	22.9	22.4	21.5	a/	9476	9499	10110	9140
052	Mining of lignite		7	5	6	6	a/	...	...	...	...	a/	...	...	...	...
061	Extraction of crude petroleum		4	4	3	3		...	...	...	...		...	...	...	...
062	Extraction of natural gas		1	1	2	2		...	...	...	...		...	...	...	...
071	Mining of iron ores		-	-	-	-		-	-	-	-		-	-	-	-
072	Mining of non-ferrous metal ores		1	1	1	1		...	...	...	...		...	...	...	...
081	Quarrying of stone, sand and clay		245	257	270	271		5.7	5.5	5.4	5.2		1788	1770	1781	1665
089	Mining and quarrying n.e.c.		40	43	37	41		0.5	0.5	0.5	0.5		177	150	166	153
091	Support activities for petroleum		7	7	5	5		0.4	0.5	0.5	2.3b/		185	182	206	746b/
099	Support activities for other mining, quarrying		23	25	30	28		1.7	1.8	1.7	...b/		468	531	546	...b/
B	Mining and quarrying		333	348	359	363		35.4	34.2	33.6	32.6		13198	13207	13993	12904
351	Electric power generation, transmission		2819	4751	5546	8000		14.3	14.6	15.3	15.8		8546	8835	9577	9584
352	Manufacture of gas		40	36	41	46		2.4	2.5	2.6	2.1		1075	1198	1335	948
353	Steam and air conditioning supply		408	405	404	400		13.3	12.8	12.4	11.7		4855	4756	4854	4484
D	Electricity, gas, steam & air conditioning		3267	5192	5991	8446		30.0	29.9	30.3	29.6		14475	14788	15766	15016
360	Water collection, treatment and supply		332	386	446	472		17.9	17.8	17.7	17.6		5568	5678	5850	5996
370	Sewerage		444	405	349	312		1.1	1.2	1.2	1.1		351	376	360	348
381	Waste collection		3556	4011	4120	4049		18.9	19.8	20.4	20.7		4973	5213	5382	5341
382	Waste treatment and disposal		482	436	406	386		3.0	3.0	2.9	2.8		920	878	890	873
383	Materials recovery		1130	1054	982	900		4.8	5.0	5.1	4.9		1178	1261	1268	1200
390	Remediation activities		90	91	80	77		0.4	0.6	0.6	0.6		154	231	231	226
E	Water supply;sewerage,waste management		6034	6383	6383	6196		46.1	47.3	47.8	47.7		13143	13640	13983	13985

a/ 051 includes 052.
b/ 091 includes 099.

Czech Republic

ISIC Revision 4		Output (valuation not defined)					Value added at factor values				
			(millions of Czech Koruny)					(millions of Czech Koruny)			
ISIC	Industry	Note	2010	2011	2012	2013	Note	2010	2011	2012	2013
051	Mining of hard coal	a/	63359	66371	61082	50578	a/	33251	35474	29638	22649
052	Mining of lignite	a/	...	...	...	...	a/	...	...	...	...
061	Extraction of crude petroleum		...	...	...	...		...	...	...	...
062	Extraction of natural gas		...	...	...	...		...	...	...	...
071	Mining of iron ores		-	-	-	-		-	-	-	-
072	Mining of non-ferrous metal ores		...	...	...	...		...	...	...	...
081	Quarrying of stone, sand and clay		14715	14658	13970	13736		5138	4780	4424	4380
089	Mining and quarrying n.e.c.		1603	1682	1728	2006		425	460	438	473
091	Support activities for petroleum		1163	3910b/	4487b/	4458b/		521	1517b/	1788b/	2065b/
099	Support activities for other mining, quarrying		2187	...b/	...b/	...b/		748	...b/	...b/	...b/
B	Mining and quarrying		87273	91263	86741	76496		43324	45927	40299	33631
351	Electric power generation, transmission		545712	650560	682300	666587		112504	114592	122194	128775
352	Manufacture of gas		213819	229818	271838	270241		22756	11857	-4102	7506
353	Steam and air conditioning supply		61354	57735	60742	58575		18538	16812	17934	17692
D	Electricity, gas, steam & air conditioning		820883	938113	1014880	995403		153798	143261	136026	153973
360	Water collection, treatment and supply		33542	34837	36464	37855		13944	14265	15034	15357
370	Sewerage		2430	2688	2447	2359		875	1057	956	977
381	Waste collection		38133	41594	40382	38762		12240	12762	12753	12616
382	Waste treatment and disposal		5691	5562	5347	5534		2240	2129	2133	2091
383	Materials recovery		23630	27762	24940	22190		3914	3838	3259	2642
390	Remediation activities		981	1161	1262	1289		301	101	103	99
E	Water supply;sewerage,waste management		104408	113606	110842	107989		33514	34153	34240	33782

a/ 051 includes 052.
b/ 091 includes 099.

Denmark

ISIC Revision 4		Number of enterprises					Number of employees					Wages and salaries paid to employees				
		Note	(number)				Note	(number)				Note	(millions of Danish Kroner)			
ISIC	Industry		2010	2011	2012	2013		2010	2011	2012	2013		2010	2011	2012	2013
051	Mining of hard coal		-	-	-	-		-	-	-	-		-	-	-	-
052	Mining of lignite		-	-	-	-		-	-	-	-		-	-	-	-
061	Extraction of crude petroleum		9	8	9	10		566	849	952	966		479	606	552	569
062	Extraction of natural gas		-	-	-	-		-	-	-	-		-	-	-	-
071	Mining of iron ores		-	-	-	-		-	-	-	-		-	-	-	-
072	Mining of non-ferrous metal ores		-	-	-	-		-	-	-	-		-	-	-	-
081	Quarrying of stone, sand and clay		139	144	136	129		641	860	825	772		238	285	269	262
089	Mining and quarrying n.e.c.		14	15	13	10		385	468	470	459		160	156	162	160
091	Support activities for petroleum		35	40	49	52		2090a/	3072a/	3434	1966		1144a/	1262a/	1583	873
099	Support activities for other mining, quarrying		6	7	11	9		...a/	...a/	77	133		...a/	...a/	33	46
B	Mining and quarrying		203	214	218	210		3682	5249	5758	4296		2021	2309	2599	1909
351	Electric power generation, transmission		1239	1367	1447	1387		8901	8715	7453	8782		3657	3543	2471	4226
352	Manufacture of gas		49	49	54	61		586	577	1851	569		534	698	1869	509
353	Steam and air conditioning supply		393	401	401	397		1388	1409	1453	1499		599	610	616	676
D	Electricity, gas, steam & air conditioning		1681	1817	1902	1845		10875	10701	10757	10850		4790	4851	4955	5410
360	Water collection, treatment and supply		1994	1979	2012	2007		2306	2510	2546	2481		392	423	345	212
370	Sewerage		231	235	241	254		1883	2137	2101	1885		798	856	851	784
381	Waste collection		107	115	113	104		3144	3055	2073	2198		1253	1235	1246	923
382	Waste treatment and disposal		63	71	69	69		1850	1888	1959	2034		790	782	809	841
383	Materials recovery		138	152	147	127		1108	1432	1698	1551		409	437	529	494
390	Remediation activities		13	14	18	18		58	52	53	80		20	28	36	7
E	Water supply;sewerage,waste management		2546	2566	2600	2579		10349	11074	10430	10229		3663	3763	3816	3261

a/ 091 includes 099.

Denmark

ISIC Revision 4		Output (valuation not defined)					Value added at factor values				
			(millions of Danish Kroner)					(millions of Danish Kroner)			
ISIC	Industry	Note	2010	2011	2012	2013	Note	2010	2011	2012	2013
051	Mining of hard coal		-	-	-	-		-	-	-	-
052	Mining of lignite		-	-	-	-		-	-	-	-
061	Extraction of crude petroleum		52539	60644	54956	51202		43602	49393	43398	38801
062	Extraction of natural gas		-	-	-	-		-	-	-	-
071	Mining of iron ores		-	-	-	-		-	-	-	-
072	Mining of non-ferrous metal ores		-	-	-	-		-	-	-	-
081	Quarrying of stone, sand and clay		1228	1856	1976	1768		499	630	630	548
089	Mining and quarrying n.e.c.		699	704	799	775		319	320	334	315
091	Support activities for petroleum		2935a/	3884a/	4697	2932		1857a/	2068a/	2650	1739
099	Support activities for other mining, quarrying		...a/	...a/	93	94		...a/	...a/	28	44
B	Mining and quarrying		57402	67089	62521	56771		46278	52411	47040	41447
351	Electric power generation, transmission		54173	59238	52346	104696		14501	13301	14788	20080
352	Manufacture of gas		40708	45116	63624	46903		6407	7162	-2696	69
353	Steam and air conditioning supply		14640	15577	16220	17816		3307	3465	3932	3531
D	Electricity, gas, steam & air conditioning		109522	119932	132190	169415		24214	23928	16023	23679
360	Water collection, treatment and supply		6033	6554	6261	6196		2190	2544	2791	2721
370	Sewerage		8410	9344	9877	11159		4222	5034	5488	6567
381	Waste collection		5308	5826	5476	4911		2067	2138	1998	1407
382	Waste treatment and disposal		5084	6016	5572	6392		1909	2328	2022	1909
383	Materials recovery		3610	6038	5825	5480		991	1071	986	842
390	Remediation activities		38	48	51	134		1	-6	-7	35
E	Water supply;sewerage,waste management		28484	33826	33060	34273		11379	13111	13278	13480

a/ 091 includes 099.

Egypt

ISIC Revision 4		Number of establishments					Number of persons engaged					Wages and salaries paid to employees				
		Note	(number)				Note	(number)				Note	(millions of Egyptian Pounds)			
ISIC	Industry		2010	2011	2012	2013		2010	2011	2012	2013 a/		2010	2011	2012	2013
05	Mining of coal and lignite		...	...	...	...		...	...	...	...		...	...	...	...
06	Extraction of crude petroleum & natural gas		...	...	21	31		...	...	22676	21532		...	...	3340.9	5810.8
07	Mining of metal ores		...	...	3	3		...	...	2282	2239		...	...	131.4	137.2
08	Other mining and quarrying		...	...	13	125		...	...	4535	9875		...	...	239.1	395.9
09	Mining support service activities		...	...	...	...		...	...	34	34		...	...	1.1	0.8
B	Mining and quarrying		...	...	37b/	159b/		...	...	29527b/	33680b/		...	...	3712.5b/	6344.7b/
35	Electricity, gas, steam & air conditioning		...	...	...	...		...	...	...	...		...	...	...	...
D	Electricity, gas, steam & air conditioning		...	...	...	...		...	...	...	...		...	...	...	...
36	Water collection, treatment and supply		...	...	...	...		...	...	...	...		...	...	...	...
37	Sewerage		...	...	...	...		...	...	...	...		...	...	...	...
38	Waste collection,treatment,disposal activities		...	...	...	...		...	...	...	...		...	...	...	...
39	Remediation activities		...	...	...	...		...	...	...	...		...	...	...	...
E	Water supply;sewerage,waste management		...	...	...	...		...	...	...	...		...	...	...	...

a/ Number of employees.
b/ Sum of available data.

Egypt

ISIC Revision 4		Output at producers' prices					Value added at producers' prices				
			(millions of Egyptian Pounds)					(millions of Egyptian Pounds)			
ISIC	Industry	Note	2010	2011	2012	2013	Note	2010	2011	2012	2013
05	Mining of coal and lignite		...	...	...	...		...	...	...	...
06	Extraction of crude petroleum & natural gas		...	...	213665.4	210202.0		...	...	2311.9	195927.0
07	Mining of metal ores		...	...	1475.0	1457.1		...	...	330.0	648.0
08	Other mining and quarrying		...	...	2648.7	2451.0		...	...	191.7	1446.0
09	Mining support service activities		...	...	...	...		...	...	-0.1	0.8
B	Mining and quarrying		...	...	217789.2a/	214110.1a/		...	...	2833.5a/	198021.9a/
35	Electricity, gas, steam & air conditioning		...	...	...	...		...	...	...	...
D	Electricity, gas, steam & air conditioning		...	...	...	...		...	...	...	...
36	Water collection, treatment and supply		...	...	...	...		...	...	...	...
37	Sewerage		...	...	...	...		...	...	...	...
38	Waste collection,treatment,disposal activities		...	...	...	...		...	...	...	...
39	Remediation activities		...	...	...	...		...	...	...	...
E	Water supply;sewerage,waste management		...	...	...	...		...	...	...	...

a/ Sum of available data.

Estonia

ISIC Revision 4		Number of enterprises					Number of employees					Wages and salaries paid to employees				
			(number)					(number)					(millions of Euros)			
ISIC	Industry	Note	2010	2011	2012	2013	Note	2010	2011	2012	2013	Note	2010	2011	2012	2013
051	Mining of hard coal		-	-	-	-		-	-	-	-		-	-	-	-
052	Mining of lignite		-	-	-	-		-	-	-	-		-	-	-	-
061	Extraction of crude petroleum		1	1	2	2		...	...	...	...		...	...	...	...
062	Extraction of natural gas		-	-	-	-		-	-	-	-		-	-	-	-
071	Mining of iron ores		-	-	-	-		-	-	-	-		-	-	-	-
072	Mining of non-ferrous metal ores		-	-	-	-		-	-	-	-		-	-	-	-
081	Quarrying of stone, sand and clay		84	68	69	83		700	712	...	...		7.3	8.7	...	...
089	Mining and quarrying n.e.c.		40	43	44	42		...	...	1058	1072		...	...	11.6	13.1
091	Support activities for petroleum		-	-	-	-		-	-	-	-		-	-	-	-
099	Support activities for other mining, quarrying		11	9	13	11		...	...	31	19		...	...	0.3	0.2
B	Mining and quarrying		136	121	128	138		4900	4980	5267	5073		58.2	64.1	70.9	74.1
351	Electric power generation, transmission		95	92	101	99		...	3502	3363	3244		...	57.6	60.9	61.5
352	Manufacture of gas		20	16	17	18		...	309	303	315		...	5.2	5.4	6.4
353	Steam and air conditioning supply		108	112	113	109		1800	1804	1725	1616		18.3	18.6	19.4	19.6
D	Electricity, gas, steam & air conditioning		223	220	231	226		5500	5615	5391	5175		76.3	81.5	85.7	87.4
360	Water collection, treatment and supply		74	75	74	77		1400	1225	1252	1196		13.9	13.5	14.4	14.6
370	Sewerage		46	34	36	45		...	...	176	...		...	...	1.8	...
381	Waste collection		55	46	54	51		1100	1158	1139	1115		10.6	11.7	12.6	13.6
382	Waste treatment and disposal		34	29	28	29		300	291	300	287		3.2	3.2	3.3	3.5
383	Materials recovery		65	77	79	82		...	496	428	421		...	3.7	3.5	3.3
390	Remediation activities		6	6	11	8		...	...	30	...		...	...	0.2	...
E	Water supply;sewerage,waste management		280	267	282	292		3500	3367	3325	3243		32.7	33.9	35.8	37.1

Estonia

ISIC Revision 4		Output (valuation not defined)					Value added at factor values				
			(millions of Euros)					(millions of Euros)			
ISIC	Industry	Note	2010	2011	2012	2013	Note	2010	2011	2012	2013
051	Mining of hard coal		-	-	-	-		-	-	-	-
052	Mining of lignite		-	-	-	-		-	-	-	-
061	Extraction of crude petroleum		...	...	...	...		...	...	...	...
062	Extraction of natural gas		-	-	-	-		-	-	-	-
071	Mining of iron ores		-	-	-	-		-	-	-	-
072	Mining of non-ferrous metal ores		-	-	-	-		-	-	-	-
081	Quarrying of stone, sand and clay		41.0	55.5	...	...		15.1	19.9	...	...
089	Mining and quarrying n.e.c.		...	...	74.6	79.6		...	...	30.1	34.3
091	Support activities for petroleum		-	-	-	-		-	-	-	-
099	Support activities for other mining, quarrying		...	...	1.4	1.2		...	...	0.7	0.6
B	Mining and quarrying		302.4	343.2	359.4	410.4		139.7	154.8	154.0	211.0
351	Electric power generation, transmission		...	675.6	677.5	606.4		...	366.5	383.2	435.8
352	Manufacture of gas		...	...	...	...		...	28.7	30.9	33.8
353	Steam and air conditioning supply		214.0	213.0	258.7	249.5		92.3	79.9	109.5	127.0
D	Electricity, gas, steam & air conditioning		898.2	888.6	936.3	856.5		445.8	475.1	523.7	596.6
360	Water collection, treatment and supply		...	...	...	...		71.5	75.7	70.4	68.2
370	Sewerage		...	...	...	...		...	...	4.5	...
381	Waste collection		...	...	...	...		24.7	29.6	28.1	31.0
382	Waste treatment and disposal		...	...	...	...		11.9	11.4	8.8	9.3
383	Materials recovery		...	...	...	...		...	9.4	10.4	1.7
390	Remediation activities		...	...	...	...		...	...	0.6	...
E	Water supply;sewerage,waste management		...	...	...	...		131.8	131.5	122.7	114.5

Finland

ISIC Revision 4		Number of enterprises					Number of employees					Wages and salaries paid to employees				
			(number)					(number)					(millions of Euros)			
ISIC	Industry	Note	2010	2011	2012	2013	Note	2010	2011	2012	2013	Note	2010	2011	2012	2013
05	Mining of coal and lignite		-	-	-	-		-	-	-	-		-	-	-	-
051	Mining of hard coal		-	-	-	-		-	-	-	-		-	-	-	-
052	Mining of lignite		-	-	-	-		-	-	-	-		-	-	-	-
06	Extraction of crude petroleum & natural gas		-	-	-	-		-	-	-	-		-	-	-	-
061	Extraction of crude petroleum		-	-	-	-		-	-	-	-		-	-	-	-
062	Extraction of natural gas		-	-	-	-		-	-	-	-		-	-	-	-
07	Mining of metal ores		26	24	24	27		1434	...	1911	2145		53	...	88	104
071	Mining of iron ores		-	-	-	1		-	-	-	2145a/		-	-	-	104a/
072	Mining of non-ferrous metal ores		26	24	24	26		1434	...	1911	...a/		53	...	88	...a/
08	Other mining and quarrying		820	820	828	823		3671	3763	3383	3473		114	118	114	111
081	Quarrying of stone, sand and clay		345	344	343	334		1776	1795	1579	1498		57	59	55	53
089	Mining and quarrying n.e.c.		475	476	485	489		1895	1968	1804	1976		57	59	59	58
09	Mining support service activities		23	40	36	45		208	...	355	368		8	...	15	15
091	Support activities for petroleum		-	-	-	-		-	-	-	-		-	-	-	-
099	Support activities for other mining, quarrying		23	40	36	45		208	...	355	368		8	...	15	15
B	Mining and quarrying		869	884	888	895		5314	5543	5649	5987		175	196	218	229
35	Electricity, gas, steam & air conditioning		736	764	783	816		13453	13458	13786	13659		624	638	665	661
351	Electric power generation, transmission		344	357	368	399		12530	12592	12411	12160		588	603	610	598
352	Manufacture of gas		10	14	11	11		...	59	...	...		...	3	...	...
353	Steam and air conditioning supply		382	393	404	406		...	808	...	...		...	32	...	...
D	Electricity, gas, steam & air conditioning		736	764	783	816		13453	13458	13786	13659		624	638	665	661
36	Water collection, treatment and supply		676	700	699	736		2737	2867	2944	1653		101	107	112	67
37	Sewerage		149	167	182	197		415	443	...	...		14	16	...	...
38	Waste collection,treatment,disposal activities		509	506	511	505		5494	...	...	...		175	...	...	...
381	Waste collection		268	256	251	245		2172	2295	2304	1833		62	66	70	61
382	Waste treatment and disposal		132	132	136	126		1974	1823	1813	1884		65	66	68	72
383	Materials recovery		109	118	124	134		1348	...	...	...		48	...	...	...
39	Remediation activities		47	46	44	41		402	...	385	...		13	...	13	...
E	Water supply;sewerage,waste management		1381	1419	1436	1479		9048	9278	9416	7618		302	318	334	280

a/ 071 includes 072.

Finland

ISIC Revision 4		Output (valuation not defined)					Value added at factor values				
			(millions of Euros)					(millions of Euros)			
ISIC	Industry	Note	2010	2011	2012	2013	Note	2010	2011	2012	2013
05	Mining of coal and lignite		-	-	-	-		-	-	-	-
051	Mining of hard coal		-	-	-	-		-	-	-	-
052	Mining of lignite		-	-	-	-		-	-	-	-
06	Extraction of crude petroleum & natural gas		-	-	-	-		-	-	-	-
061	Extraction of crude petroleum		-	-	-	-		-	-	-	-
062	Extraction of natural gas		-	-	-	-		-	-	-	-
07	Mining of metal ores		538	...	786	685		257	...	299	170
071	Mining of iron ores		-	-	-	685a/		-	-	-	170a/
072	Mining of non-ferrous metal ores		538	...	786	...a/		257	...	299	...a/
08	Other mining and quarrying		1099	1180	1047	998		334	307	282	295
081	Quarrying of stone, sand and clay		419	483	424	376		121	131	120	111
089	Mining and quarrying n.e.c.		679	698	623	622		213	176	162	185
09	Mining support service activities		29	...	54	40		11	...	-	2
091	Support activities for petroleum		-	-	-	-		-	-	-	-
099	Support activities for other mining, quarrying		29	...	54	40		11	...	-	2
B	Mining and quarrying		1666	2483	1887	1723		602	1200	580	467
35	Electricity, gas, steam & air conditioning		7463	7848	7916	7861		3608	3313	3355	3481
351	Electric power generation, transmission		7156	7414	7304	6793		3431	3159	3108	3193
352	Manufacture of gas		...	24	...	...		...	12	...	...
353	Steam and air conditioning supply		...	410	...	...		...	143	...	...
D	Electricity, gas, steam & air conditioning		7463	7848	7916	7861		3608	3313	3355	3481
36	Water collection, treatment and supply		803	861	915	516		465	494	510	277
37	Sewerage		78	87	...	...		46	49	...	...
38	Waste collection,treatment,disposal activities		1429	...	...	...		404	...	...	...
381	Waste collection		299	322	332	306		123	126	126	113
382	Waste treatment and disposal		531	581	616	648		164	172	187	198
383	Materials recovery		599	...	...	...		118	...	...	...
39	Remediation activities		77	...	82	...		27	...	30	...
E	Water supply;sewerage,waste management		2387	2628	2754	2273		942	996	1013	758

a/ 071 includes 072.

France

ISIC Revision 4		Number of enterprises					Number of employees					Wages and salaries paid to employees				
			(number)					(number)					(millions of Euros)			
ISIC	Industry	Note	2010	2011	2012	2013	Note	2010	2011	2012	2013	Note	2010	2011	2012	2013
05	Mining of coal and lignite		6	-	-	...		28	-	-	...		...	-	-	...
051	Mining of hard coal		6	-	-	...		28	-	-	...		...	-	-	...
052	Mining of lignite		-	-	-	-		-	-	-	-		-	-	-	-
06	Extraction of crude petroleum & natural gas		32	57	57	58		814	964	692	672		49	...	46	43
061	Extraction of crude petroleum		32a/	57a/	50	58a/	a/	814	964	692	672		49a/	...	46a/	43a/
062	Extraction of natural gas		...a/	...a/	7	...a/	a/	...	...	...	...		...a/	...	...a/	...a/
07	Mining of metal ores		54	48	56	...		352	393	...	...		...	...	...	...
071	Mining of iron ores		3	3	56b/	...		352b/	393b/	...	...		...	...	...	...
072	Mining of non-ferrous metal ores		51	45	...b/	84		...b/	...b/	...	498		...	...	...	...
08	Other mining and quarrying		1673	1630	1543	1637		24762	23690	22780	22329		777	789	770	774
081	Quarrying of stone, sand and clay		1520	1454	1403	1462		23037	22039	21129	20581		707	718	701	702
089	Mining and quarrying n.e.c.		153	176	141	175		1724	1651	1651	1748		70	71	70	73
09	Mining support service activities		55	57	62	80		189	...	...	367		43	...	...	52
091	Support activities for petroleum		36	35	39	45		110	156	...	367c/		39	42	...	52c/
099	Support activities for other mining, quarrying		19	22	23	35		79	...	...	...c/		4	...	...	...c/
B	Mining and quarrying		1820	1792	1718	1861		26145	25245	24246	23873		924	905	877	887
35	Electricity, gas, steam & air conditioning		14337	16776	18554	20756		169639	171276	175905	177437		7784	7997	8142	8418
351	Electric power generation, transmission		13932	16311	18113	20254		127083	129980	134305	136229		5763	6032	6237	6500
352	Manufacture of gas		78	122	93	104		18212	...	17225	16871		1126	1051	963	952
353	Steam and air conditioning supply		328	343	347	398		24344	...	24376	24337		895	914	943	965
D	Electricity, gas, steam & air conditioning		14337	16776	18554	20756		169639	171276	175905	177437		7784	7997	8142	8418
36	Water collection, treatment and supply		...	3049	3018	3038		41943	42614	41623	40177		1366	1526	1491	1476
37	Sewerage		1708	1671	1718	1774		13882	13141	15148	15447		461	451	473	505
38	Waste collection,treatment,disposal activities		7250	6708	8322	8686		91435	92338	95063	91664		2737	2779	2978	2898
381	Waste collection		932	1124	1157	1354		40781	41659	41540	40366		1164	1210	1262	1246
382	Waste treatment and disposal		713	690	1078	820		19821	21490	22246	20343		667	703	751	711
383	Materials recovery		5605	4894	6087	6512		30833	29189	31277	30954		906	867	966	941
39	Remediation activities		200	231	438	322		3683	4064	4989	6388		116	121	167	208
E	Water supply;sewerage,waste management		12205	11659	13497	13820		150943	152157	156823	153676		4680	4877	5109	5087

a/ 061 includes 062.
b/ 071 includes 072.
c/ 091 includes 099.

France

ISIC Revision 4		Output (valuation not defined)					Value added at factor values				
		Note	(millions of Euros)				Note	(millions of Euros)			
ISIC	Industry		2010	2011	2012	2013		2010a/	2011	2012	2013
05	Mining of coal and lignite		...	-	-	...		...	-	-	...
051	Mining of hard coal		...	-	-	...		...	-	-	...
052	Mining of lignite		-	-	-	-		-	-	-	-
06	Extraction of crude petroleum & natural gas		742	...	1004	959		367	...	408	280
061	Extraction of crude petroleum		742b/	...	1004b/	959b/		367b/	...	408b/	280b/
062	Extraction of natural gas		...b/	...	...b/	...b/		...b/	...	...b/	...b/
07	Mining of metal ores		...	...	...	...		...	...	...	...
071	Mining of iron ores		-	...	...	...		-	...	...	...
072	Mining of non-ferrous metal ores		...	...	...	...		...	...	...	...
08	Other mining and quarrying		6025	6399	6180	6403		2036	...	1913	1957
081	Quarrying of stone, sand and clay		5543	5886	5750	5894		1876	1901	1766	1786
089	Mining and quarrying n.e.c.		481	513	429	508		160	...	147	172
09	Mining support service activities		281	...	...	121		74	...	...	-79
091	Support activities for petroleum		265	243	...	121c/		72	109	...	-79c/
099	Support activities for other mining, quarrying		16	...	...	...c/		2	...	...	...c/
B	Mining and quarrying		7197	7665	7434	7576		2478	2582	2316	2175
35	Electricity, gas, steam & air conditioning		112558	113740	122255	122083		24866	24704	28822	31131
351	Electric power generation, transmission		72164	74991	78706	77991		19579	20827	24272	25986
352	Manufacture of gas		33942	32019	36084	36450		3731	2267	2840	3435
353	Steam and air conditioning supply		6451	6730	7465	7642		1557	1611	1710	1710
D	Electricity, gas, steam & air conditioning		112558	113740	122255	122083		24866	24704	28822	31131
36	Water collection, treatment and supply		12843	12574	12390	12091		3464	3629	3461	3643
37	Sewerage		2490	2548	2525	2804		1214	1214	1204	1331
38	Waste collection,treatment,disposal activities		20401	22279	22995	21080		5719	5763	5762	5303
381	Waste collection		4883	4952	5315	5224		1955	2006	2140	2123
382	Waste treatment and disposal		5073	5143	5485	5261		1688	1782	1758	1560
383	Materials recovery		10446	12184	12196	10594		2076	1975	1865	1620
39	Remediation activities		620	689	893	1035		219	229	286	366
E	Water supply;sewerage,waste management		36354	38089	38803	37010		10616	10836	10713	10642

a/ Sum of available data.
b/ 061 includes 062.
c/ 091 includes 099.

Georgia

ISIC Revision 3		Number of establishments					Number of employees					Wages and salaries paid to employees				
			(number)					(number)					(thousands of Georgian Lari)			
ISIC	Industry	Note	2010	2011	2012	2013	Note	2010	2011	2012	2013	Note	2010	2011	2012	2013
10	Mining of coal and lignite; extraction of peat		...	...	...	...		...	...	...	...		...	...	...	...
101	Mining and agglomeration of hard coal		1	4	10	16		1022	1344	1690	1549		7561	13165	16468	14161
102	Mining and agglomeration of lignite		...	...	...	...		...	...	...	...		...	...	...	...
103	Extraction and agglomeration of peat		1	1	1	1		...	...	...	...		...	...	...	...
11	Extraction of crude oil and natural gas, etc.		9	23	18	17		...	...	...	...		...	...	...	...
111	Extraction of crude oil and natural gas		7	15	11	13		576	787	915	969		12713	14901	15867	17769
112	Oil, gas service activities		2	8	7	4		...	...	...	...		...	...	...	...
12	Mining of uranium and thorium ores		...	...	...	...		...	...	...	...		...	...	...	...
120	Mining of uranium and thorium ores		...	...	...	...		...	...	...	...		...	...	...	...
13	Mining of metal ores		...	...	...	...		...	...	...	...		...	...	...	...
131	Iron ores		...	...	...	...		...	...	...	...		...	...	...	...
132	Non-ferrous metal ores,excl.uranium,thorium		10	8	11	16		1263	1144	1219	1345		17220	15810	19979	21309
14	Other mining and quarrying		223	298	323	349		2115	2431	3037	2415		11046	13606	18324	14103
141	Quarrying of stone, sand and clay		213	281	315	340		2040	2335	2932	2312		10513	12841	17503	13449
142	Mining and quarrying n.e.c.		10	17	8	9		75	96	105	103		533	765	821	654
C	Total mining and quarrying		244	334	363	399		4981	5714	6918	6289		48556	57503	72597	67400
40	Electricity, gas, steam and hot water supply		92	122	108	105		15434	16353	16642	16557		168851	185542	205059	209973
401	Electricity production, collection, distribution		42	60	49	55		11327	11650	12302	12566		138410	147964	167585	172251
402	Gas production and distribution		46	57	55	48		4043	4621	4330	3967		30124	37184	37406	37572
403	Steam and hot water supply		4	5	4	2		64	82	10	24		317	394	68	150
41	Water collection, purification, distribution		17	21	16	17		4928	5099	5228	5226		32217	40229	38784	46048
410	Water collection, purification, distribution		17	21	16	17		4928	5099	5228	5226		32217	40229	38784	46048
E	Total electricity, gas and water supply		109	143	124	122		20362	21452	21870	21783		201069	225771	243844	256021

Georgia

ISIC Revision 3		Output (valuation not defined)					Value added (valuation not defined)				
			(thousands of Georgian Lari)					(thousands of Georgian Lari)			
ISIC	Industry	Note	2010	2011	2012	2013	Note	2010	2011	2012	2013
10	Mining of coal and lignite; extraction of peat		...	...	...	...		...	...	...	...
101	Mining and agglomeration of hard coal		27543	40019	50951	41894		12804	21755	39954	26173
102	Mining and agglomeration of lignite		...	...	...	...		...	...	...	...
103	Extraction and agglomeration of peat		...	...	...	...		...	...	...	...
11	Extraction of crude oil and natural gas, etc.		...	...	...	...		...	...	...	...
111	Extraction of crude oil and natural gas		49706	67802	72164	78464		33194	38202	23119	30436
112	Oil, gas service activities		...	...	...	...		...	...	...	...
12	Mining of uranium and thorium ores		...	...	...	...		...	...	...	...
120	Mining of uranium and thorium ores		...	...	...	...		...	...	...	...
13	Mining of metal ores		...	...	...	...		...	...	...	...
131	Iron ores		...	...	...	...		...	...	...	...
132	Non-ferrous metal ores,excl.uranium,thorium		125724	141289	118107	128984		69016	67171	60738	59406
14	Other mining and quarrying		51947	79981	119630	75906		26963	40551	53232	37024
141	Quarrying of stone, sand and clay		48907	73655	115697	72725		25323	37397	51790	35565
142	Mining and quarrying n.e.c.		3040	6326	3933	3181		1640	3154	1442	1459
C	Total mining and quarrying		254947	329143	364390	325332		141997	167708	178643	153104
40	Electricity, gas, steam and hot water supply		728923	888278	895642	955596		523277	541774	475921	581910
401	Electricity production, collection, distribution		619528	668581	770814	811341		479411	469964	413480	501746
402	Gas production and distribution		108651	218475	124388	143782		43225	70753	62023	79782
403	Steam and hot water supply		744	1222	440	473		641	1057	418	382
41	Water collection, purification, distribution		142079	136585	163062	153637		93739	80546	108139	102625
410	Water collection, purification, distribution		142079	136585	163062	153637		93739	80546	108139	102625
E	Total electricity, gas and water supply		871002	1024864	1058730	1109233		617016	622320	584060	684535

Germany

ISIC Revision 4		Number of enterprises					Number of employees					Wages and salaries paid to employees				
			(number)					(thousands)					(millions of Euros)			
ISIC	Industry	Note	2010	2011	2012	2013	Note	2010	2011	2012	2013	Note	2010	2011	2012	2013
051	Mining of hard coal	a/	6	5	5	5	a/	33.7	28.9	27.0	22.5	a/	1295	1147	1100	968
052	Mining of lignite	a/	...	...	...	...	a/	...	...	...	...	a/	...	...	...	...
061	Extraction of crude petroleum		-	-	-	-		-	-	-	-		-	-	-	-
062	Extraction of natural gas		4	4	4	4		3.8	3.8	3.8	3.8		339	344	363	369
071	Mining of iron ores		-	-	-	-		-	-	-	-		-	-	-	-
072	Mining of non-ferrous metal ores		-	-	-	-		-	-	-	-		-	-	-	-
081	Quarrying of stone, sand and clay		1581	1476	1593	1450		25.7	26.4	25.4	25.0		819	862	831	845
089	Mining and quarrying n.e.c.		165	177	166	155		5.1	5.1	5.0	5.0		212	218	208	211
091	Support activities for petroleum	b/	80	105	133	92	b/	2.8	3.0	3.5	3.4	b/	132	144	167	164
099	Support activities for other mining, quarrying	b/	...	...	...	...	b/	...	...	...	...	b/	...	...	...	...
B	Mining and quarrying		1836	1767	1902	1706		71.1	67.2	64.7	59.7		2797	2714	2669	2557
351	Electric power generation, transmission		1205	1240	1344	1402		...	223.9c/	222.3c/	219.9c/		10389	10848	10843	10989
352	Manufacture of gas		263	270	284	282		16.8	...c/	...c/	...c/		898	890	875	781
353	Steam and air conditioning supply		254	267	271	290		...	...c/	...c/	...c/		526	582	713	718
D	Electricity, gas, steam & air conditioning		1722	1777	1899	1974		221.2	223.9	222.3	219.9		11814	12321	12432	12487
360	Water collection, treatment and supply		1660	1672	1631	1646		...	...	...	...		1292	1365	1357	1379
370	Sewerage		1318	1370	1404	1415		...	...	...	...		1485	1577	1656	1729
381	Waste collection		622	703	765	755		50.7	57.2	61.4	63.1		1499	1742	1890	1967
382	Waste treatment and disposal		501	540	568	559		35.4	36.5	37.8	36.9		1229	1289	1355	1373
383	Materials recovery		455	554	595	605		20.2	23.9	24.1	24.3		576	676	726	749
390	Remediation activities		59	59	64	66		...	...	...	...		86	97	99	103
E	Water supply;sewerage,waste management		4615	4898	5027	5046		183.5	198.6	204.2	206.3		6167	6746	7083	7300

a/ 051 includes 052.
b/ 091 includes 099.
c/ 351 includes 352 and 353.

Germany

ISIC Revision 4		Output (valuation not defined)					Value added at factor values				
			(millions of Euros)					(millions of Euros)			
ISIC	Industry	Note	2010	2011	2012	2013	Note	2010	2011	2012	2013
051	Mining of hard coal	a/	3900	3173	2829	2354	a/	3252	2033	2091	1785
052	Mining of lignite	a/	...	...	...	...	a/	...	...	...	...
061	Extraction of crude petroleum		-	-	-	-		-	-	-	-
062	Extraction of natural gas		2498	2850	2997	2855		1168	1192	1407	1159
071	Mining of iron ores		-	-	-	-		-	-	-	-
072	Mining of non-ferrous metal ores		-	-	-	-		-	-	-	-
081	Quarrying of stone, sand and clay		4210	4875	4861	4621		1695	1889	1926	1797
089	Mining and quarrying n.e.c.		1330	1335	1211	1360		509	497	462	505
091	Support activities for petroleum	b/	469	542	559	496	b/	211	233	272	241
099	Support activities for other mining, quarrying	b/	...	...	...	...	b/	...	...	...	...
B	Mining and quarrying		12407	12775	12456	11686		6835	5844	6157	5487
351	Electric power generation, transmission		...	...	...	...		...	36304	39467	37292
352	Manufacture of gas		...	...	...	...		...	4235	5531	3653
353	Steam and air conditioning supply		...	...	...	...		...	1682	2549	2358
D	Electricity, gas, steam & air conditioning		...	...	...	...		...	42221	47547	43302
360	Water collection, treatment and supply		...	...	...	...		...	4544	4370	4486
370	Sewerage		...	...	...	...		...	6876	7005	7163
381	Waste collection		...	...	...	...		...	3721	3946	4015
382	Waste treatment and disposal		...	...	...	...		...	3492	3412	3439
383	Materials recovery		...	...	...	...		...	2230	1961	1853
390	Remediation activities		...	...	...	...		...	180	157	187
E	Water supply;sewerage,waste management		...	...	...	...		...	21042	20851	21143

a/ 051 includes 052.
b/ 091 includes 099.

Greece

ISIC Revision 4		Number of enterprises					Number of employees					Wages and salaries paid to employees				
			(number)					(number)					(millions of Euros)			
ISIC	Industry	Note	2010	2011	2012	2013	Note	2010	2011	2012	2013	Note	2010	2011	2012	2013
051	Mining of hard coal		-	...	...	-		-	...	...	-		-	...	...	-
052	Mining of lignite		...	...	...	...		...	...	...	...		...	...	...	...
061	Extraction of crude petroleum		...	...	...	...		...	...	...	...		...	...	...	...
062	Extraction of natural gas		-	...	...	-		-	...	...	-		-	...	...	-
071	Mining of iron ores		-	...	...	-		-	...	...	-		-	...	...	-
072	Mining of non-ferrous metal ores		...	...	...	4		...	...	...	1630		...	...	...	28
081	Quarrying of stone, sand and clay		444	397a/	...	353		4176	4302a/	...	2852		107	114a/	...	62
089	Mining and quarrying n.e.c.		...	...a/	...	9		...	...a/	...	449		...	...a/	...	12
091	Support activities for petroleum		...	...	...	...		...	...	...	...		...	...	...	...
099	Support activities for other mining, quarrying		-	...	...	...		-	...	...	...		-	...	...	...
B	Mining and quarrying		463	406	380	372		5525	5145	4467	5417		150	142	119	119
351	Electric power generation, transmission		10b/	10b/	12	28		22801b/	21778b/	20317	19820		785b/	681b/	645	639
352	Manufacture of gas		...b/	...b/	...	...		...b/	...b/	...	...		...b/	...b/	...	...
353	Steam and air conditioning supply		...b/	...b/	...	...		...b/	...b/	...	...		...b/	...b/	...	...
D	Electricity, gas, steam & air conditioning		10	10	18	32		22801	21778	21441	20450		785	681	687	659
360	Water collection, treatment and supply		80	129	129	128		7572	9028	8779	8864		268	276	237	213
370	Sewerage		...	...	...	...		...	...	...	...		...	...	...	...
381	Waste collection		...	6	...	11		...	124	...	330		...	2	...	6
382	Waste treatment and disposal		...	...	4	6		...	...	396	426		...	...	10	9
383	Materials recovery		13	12	12	16		278	206	369	482		5	4	5	6
390	Remediation activities		...	...	...	...		...	...	...	...		...	...	...	...
E	Water supply;sewerage,waste management		104	154	153	166		8223	9879	9722	10202		282	294	256	236

a/ 081 includes 089.
b/ 351 includes 352 and 353.

Greece

ISIC Revision 4		Output (valuation not defined)					Value added at factor values				
			(millions of Euros)					(millions of Euros)			
ISIC	Industry	Note	2010	2011	2012	2013	Note	2010	2011	2012	2013
051	Mining of hard coal		-	...	...	-		-	...	...	-
052	Mining of lignite		...	...	...	...		...	...	...	...
061	Extraction of crude petroleum		...	...	...	...		...	...	...	...
062	Extraction of natural gas		-	...	...	-		-	...	...	-
071	Mining of iron ores		-	...	...	-		-	...	...	-
072	Mining of non-ferrous metal ores		...	...	...	141		...	...	...	58
081	Quarrying of stone, sand and clay		618	567a/	...	341		272	265a/	...	170
089	Mining and quarrying n.e.c.		...	...a/	...	66		...	...a/	...	33
091	Support activities for petroleum		...	...	...	...		...	...	...	...
099	Support activities for other mining, quarrying		-	...	...	...		-	...	...	...
B	Mining and quarrying		889	762	622	673		358	329	255	299
351	Electric power generation, transmission		5312b/	4729b/	8711	9643		2950b/	2167b/	2155	1998
352	Manufacture of gas		...b/	...b/	...	...		...b/	...b/	...	...
353	Steam and air conditioning supply		...b/	...b/	...	...		...b/	...b/	...	...
D	Electricity, gas, steam & air conditioning		5312	4729	9310	9804		2950	2167	2702	2133
360	Water collection, treatment and supply		708	787	820	973		521	610	610	791
370	Sewerage		...	...	...	...		...	...	...	...
381	Waste collection		...	12	...	35		...	7	...	14
382	Waste treatment and disposal		...	...	51	...		...	...	20	21
383	Materials recovery		63	69	71	73		21	14	24	17
390	Remediation activities		...	...	...	...		...	...	...	...
E	Water supply;sewerage,waste management		811	923	961	1110		563	653	665	844

a/ 081 includes 089.
b/ 351 includes 352 and 353.

Hungary

ISIC Revision 4		Number of enterprises					Number of employees					Wages and salaries paid to employees				
			(number)					(number)					(millions of Hungarian Forints)			
ISIC	Industry	Note	2010	2011	2012	2013	Note	2010	2011	2012	2013	Note	2010	2011	2012	2013
051	Mining of hard coal		4	4	8	7		10	12	16	15		83	112	116	59
052	Mining of lignite		5	9	9	10		100	105	194	223		275	196	492	445
061	Extraction of crude petroleum		3	4	5	4		...	-	1	9		-	-	-	297
062	Extraction of natural gas		10	6	6	6		70	80	78	69		909	1006	1012	891
071	Mining of iron ores		9a/	-	10a/	...		95a/	-	98a/	...		303a/	-	347a/	...
072	Mining of non-ferrous metal ores		...a/	7	...a/	...		...a/	102	...a/	...		...a/	363	...a/	...
081	Quarrying of stone, sand and clay		348	343	319	328		2919	2603	2356	2407		7245	6733	6450	6709
089	Mining and quarrying n.e.c.		26	25	29	26		151	150	140	124		275	223	202	208
091	Support activities for petroleum		40	34	36	35		1072	1203	1232	1176		5317	6733	7405	7778
099	Support activities for other mining, quarrying		24	27	26	23		137	136	111	63		275	223	202	119
B	Mining and quarrying		469	459	448	448		4554	4391	4226	4170		14683	15645	16285	16833
351	Electric power generation, transmission		389	420	416	421		15016	14342	14348	13903		98815	98338	99328	99095
352	Manufacture of gas		51	58	62	58		4199	4078	4008	3985		23581	23635	22619	22740
353	Steam and air conditioning supply		171	184	192	196		6087	6797	6419	6262		23306	25562	24586	25739
D	Electricity, gas, steam & air conditioning		611	662	670	675		25302	25217	24775	24150		145701	147535	146534	147574
360	Water collection, treatment and supply		331	339	334	306		19641	20012	20102	21014		58402	60791	62941	65222
370	Sewerage		474	476	450	410		3239	3178	3188	2846		9008	9331	9892	8965
381	Waste collection		485	494	495	458		7602	7713	7357	5912		16474	17740	17557	14131
382	Waste treatment and disposal		206	216	224	228		7704	6966	7485	6483		23746	23048	25483	22948
383	Materials recovery		367	348	348	336		2402	2198	2291	2292		4793	4414	4773	4958
390	Remediation activities		131	139	130	125		848	747	667	614		3003	2933	2690	2019
E	Water supply;sewerage,waste management		1994	2012	1981	1863		41436	40814	41090	39161		115426	118285	123336	118243

a/ 071 includes 072.

Hungary

ISIC Revision 4		Output (valuation not defined)					Value added at factor values				
			(millions of Hungarian Forints)					(millions of Hungarian Forints)			
ISIC	Industry	Note	2010	2011	2012	2013	Note	2010	2011	2012	2013
051	Mining of hard coal		-	-	116	119		-854	-670	-289	-119
052	Mining of lignite		1625	2794	2285	2791		744	810	752	950
061	Extraction of crude petroleum		-	-	810	17783		-	-	260	4928
062	Extraction of natural gas		21515	28328	29677	22295		12782	17097	17905	14873
071	Mining of iron ores		1295a/	-	1331a/	...		523a/	-	579a/	...
072	Mining of non-ferrous metal ores		...a/	1564	...a/	...		...a/	866	...a/	...
081	Quarrying of stone, sand and clay		46143	40285	38210	45659		18953	17070	15996	20217
089	Mining and quarrying n.e.c.		1074	1034	1070	950		441	419	231	327
091	Support activities for petroleum		21653	31234	32541	32804		8843	16790	17181	17723
099	Support activities for other mining, quarrying		1102	1117	868	742		358	615	318	327
B	Mining and quarrying		94407	106356	106907	124121		41790	52996	52962	59671
351	Electric power generation, transmission		1134868	1117033	1151215	1060984		519583	466771	505059	477753
352	Manufacture of gas		319805	362706	346174	243790		128622	121777	118506	86597
353	Steam and air conditioning supply		264020	265346	254945	242187		66060	54896	26611	16565
D	Electricity, gas, steam & air conditioning		1718693	1745085	1752334	1546960		714265	643473	650176	580915
360	Water collection, treatment and supply		212671	220842	228594	232924		108594	111972	115584	113404
370	Sewerage		54876	65875	66585	60027		31653	31820	32570	29301
381	Waste collection		75619	80626	79717	67657		36060	39726	36908	30340
382	Waste treatment and disposal		100963	115212	126460	112365		55179	53527	54177	47855
383	Materials recovery		45978	44587	45065	50409		13802	13801	11194	11815
390	Remediation activities		18430	38022	25946	18287		6529	12041	7087	8134
E	Water supply;sewerage,waste management		508536	565166	572368	541669		251816	262915	257548	240851

a/ 071 includes 072.

Iran (Islamic Republic of)

ISIC Revision 4		Number of establishments					Number of employees					Wages and salaries paid to employees				
			(number)					(number)					(billions of Iranian Rials)			
ISIC	Industry	Note	2010	2011	2012	2013	Note	2010	2011	2012	2013	Note	2010	2011	2012	2013
05	Mining of coal and lignite		...	...	96	101		...	...	12102	12257		...	...	1421	1583
051	Mining of hard coal		...	...	96	101		...	...	12102	12257		...	...	1421	1583
052	Mining of lignite		...	...	-	-		...	...	...	...		...	...	...	-
06	Extraction of crude petroleum & natural gas		...	...	...	...		...	...	...	...		...	...	...	...
061	Extraction of crude petroleum		...	...	...	...		...	...	...	...		...	...	...	...
062	Extraction of natural gas		...	...	...	...		...	...	...	...		...	...	...	...
07	Mining of metal ores		...	...	243	306		...	...	27352	32895		...	...	3820	5801
071	Mining of iron ores		...	...	122	159		...	...	12102	18396		...	...	2198	3109
072	Mining of non-ferrous metal ores		...	...	121	147		...	...	15250	14499		...	...	1622	2692
08	Other mining and quarrying		...	...	4977	5038		...	...	42204	45085		...	...	3616	4679
081	Quarrying of stone, sand and clay		...	...	4443	4463		...	...	38039	40148		...	...	3293	4165
089	Mining and quarrying n.e.c.		...	...	533	575		...	...	4165	4937		...	...	323	514
09	Mining support service activities		...	...	...	...		...	...	...	...		...	...	...	...
091	Support activities for petroleum		...	...	...	...		...	...	...	...		...	...	...	...
099	Support activities for other mining, quarrying		...	...	...	...		...	...	...	...		...	...	...	...
B	Mining and quarrying		...	...	5316	5445		...	...	81176	90236		...	...	8858	12064
35	Electricity, gas, steam & air conditioning		...	...	...	...		...	...	...	...		...	...	...	...
351	Electric power generation, transmission		...	...	...	...		...	...	...	...		...	...	...	...
352	Manufacture of gas		...	...	...	...		...	...	...	...		...	...	...	...
353	Steam and air conditioning supply		...	...	...	...		...	...	...	...		...	...	...	...
D	Electricity, gas, steam & air conditioning		...	...	...	...		...	...	...	...		...	...	...	...
36	Water collection, treatment and supply		...	...	...	...		...	...	...	...		...	...	...	...
37	Sewerage		...	...	...	...		...	...	...	...		...	...	...	...
38	Waste collection,treatment,disposal activities		...	...	...	...		...	...	...	...		...	...	...	...
381	Waste collection		...	...	...	...		...	...	...	...		...	...	...	...
382	Waste treatment and disposal		...	...	...	...		...	...	...	...		...	...	...	...
383	Materials recovery		...	...	...	...		...	...	...	...		...	...	...	...
39	Remediation activities		...	...	...	...		...	...	...	...		...	...	...	...
E	Water supply;sewerage,waste management		...	...	...	...		...	...	...	...		...	...	...	...

Iran (Islamic Republic of)

ISIC Revision 4		Output at producers' prices					Value added at producers' prices				
			(billions of Iranian Rials)					(billions of Iranian Rials)			
ISIC	Industry	Note	2010	2011	2012	2013	Note	2010	2011	2012	2013
05	Mining of coal and lignite		...	...	3997.4	4327.2		...	...	3317.7	2688.5
051	Mining of hard coal		...	...	3997.4	4327.2		...	...	3317.7	2688.5
052	Mining of lignite		...	...	...	-		...	...	...	...
06	Extraction of crude petroleum & natural gas		...	...	...	...		...	...	...	...
061	Extraction of crude petroleum		...	...	...	...		...	...	...	...
062	Extraction of natural gas		...	...	...	...		...	...	...	...
07	Mining of metal ores		...	...	78810.8	99606.9		...	...	61979.4	72525.5
071	Mining of iron ores		...	...	51695.5	74049.7		...	...	38740.7	52610.9
072	Mining of non-ferrous metal ores		...	...	27115.3	25557.2		...	...	23238.7	19914.6
08	Other mining and quarrying		...	...	19573.3	25408.6		...	...	...	18218.1
081	Quarrying of stone, sand and clay		...	...	17453.2	22624.0		...	...	...	16166.7
089	Mining and quarrying n.e.c.		...	...	2120.1	2784.5		...	...	...	2051.4
09	Mining support service activities		...	...	...	...		...	...	...	...
091	Support activities for petroleum		...	...	...	...		...	...	...	...
099	Support activities for other mining, quarrying		...	...	...	...		...	...	...	...
B	Mining and quarrying		...	...	102381.4	129342.6		...	...	79163.5	93432.2
35	Electricity, gas, steam & air conditioning		...	...	...	...		...	...	...	...
351	Electric power generation, transmission		...	...	...	...		...	...	...	...
352	Manufacture of gas		...	...	...	...		...	...	...	...
353	Steam and air conditioning supply		...	...	...	...		...	...	...	...
D	Electricity, gas, steam & air conditioning		...	...	...	...		...	...	...	...
36	Water collection, treatment and supply		...	...	...	...		...	...	...	...
37	Sewerage		...	...	...	...		...	...	...	...
38	Waste collection,treatment,disposal activities		...	...	...	...		...	...	...	...
381	Waste collection		...	...	...	...		...	...	...	...
382	Waste treatment and disposal		...	...	...	...		...	...	...	...
383	Materials recovery		...	...	...	...		...	...	...	...
39	Remediation activities		...	...	...	...		...	...	...	...
E	Water supply;sewerage,waste management		...	...	...	...		...	...	...	...

Ireland

ISIC Revision 4		Number of enterprises					Number of employees					Wages and salaries paid to employees				
			(number)					(number)					(millions of Euros)			
ISIC	Industry	Note	2010	2011	2012	2013 a/	Note	2010	2011	2012	2013	Note	2010	2011	2012	2013 a/
051	Mining of hard coal		...	...	...	...		...	...	...	...		...	...	...	...
052	Mining of lignite		-	-	-	-		-	-	-	...		-	-	-	-
061	Extraction of crude petroleum		-	-	-	-		-	-	-	...		-	-	-	-
062	Extraction of natural gas		...	...	...	...		...	...	...	...		...	...	...	...
071	Mining of iron ores		-	-	-	-		-	-	-	...		-	-	-	-
072	Mining of non-ferrous metal ores		...	...	...	...		...	...	...	...		...	...	...	...
081	Quarrying of stone, sand and clay		...	108	98b/	...		...	1890	2125b/	...		...	83	85b/	...
089	Mining and quarrying n.e.c.		...	15	...b/	...		...	1583	...b/	...		...	69	...b/	...
091	Support activities for petroleum		-	6c/	-	-		-	14c/	-	...		-	1c/	-	-
099	Support activities for other mining, quarrying		...	...c/	...	...		...	...c/	...	...		...	...c/	...	...
B	Mining and quarrying		133	135	109	107		4579	4767	3436	...		242	239	182	173
351	Electric power generation, transmission		...	...	...	...		...	...	...	...		...	...	...	...
352	Manufacture of gas		...	...	...	...		...	...	...	...		...	...	...	...
353	Steam and air conditioning supply		-	-	-	-		-	-	-	...		-	-	-	-
D	Electricity, gas, steam & air conditioning		236	191	211	231		9116	9326	10486	...		657	680	705	665
360	Water collection, treatment and supply		11	11	7	...		39	48	28	...		1	2	1	...
370	Sewerage		38	47	43	37		606	711	670	...		24	29	26	26
381	Waste collection		76	82	76	77		1862	1957	1370	...		65	68	49	52
382	Waste treatment and disposal		45	52	49	47		1850	2035	1784	...		68	72	64	61
383	Materials recovery		77	91	83	95		1735	1927	1637	...		60	62	50	52
390	Remediation activities		5	8	4	7		72	66	19	...		2	2	1	1
E	Water supply;sewerage,waste management		252	291	262	280		6164	6744	5508	...		221	235	190	193

a/ Provisional data.
b/ 081 includes 089.
c/ 091 includes 099.

Ireland

ISIC Revision 4		Output (valuation not defined)					Value added at factor values				
		Note	(millions of Euros)				Note	(millions of Euros)			
ISIC	Industry		2010	2011	2012	2013a/		2010	2011	2012	2013
051	Mining of hard coal		...	...	...	...		...	...	...	...
052	Mining of lignite		-	-	-	-		-	-	-	...
061	Extraction of crude petroleum		-	-	-	-		-	-	-	...
062	Extraction of natural gas		...	...	...	...		...	...	...	...
071	Mining of iron ores		-	-	-	-		-	-	-	...
072	Mining of non-ferrous metal ores		...	...	...	...		...	...	...	...
081	Quarrying of stone, sand and clay		...	432	370b/	...		...	44	131b/	...
089	Mining and quarrying n.e.c.		...	239	...b/	...		...	116	...b/	...
091	Support activities for petroleum		-	2c/	-	-		-	1c/	-	...
099	Support activities for other mining, quarrying		...	...c/	...	...		...	...c/	...	...
B	Mining and quarrying		1084	1216	894	810		475	480	406	...
351	Electric power generation, transmission		...	...	...	...		...	...	...	...
352	Manufacture of gas		...	...	...	...		...	...	...	...
353	Steam and air conditioning supply		-	-	-	-		-	-	-	...
D	Electricity, gas, steam & air conditioning		6060	6529	6574	6550		2973	3163	3385	...
360	Water collection, treatment and supply		10	11	5	...		3	3	2	...
370	Sewerage		143	164	155	149		38	65	37	...
381	Waste collection		315	321	263	270		119	126	93	...
382	Waste treatment and disposal		339	355	338	290		128	154	108	...
383	Materials recovery		364	408	347	401		109	116	89	...
390	Remediation activities		8	7	5	5		5	4	3	...
E	Water supply;sewerage,waste management		1179	1266	1113	1127		402	469	331	...

a/ Provisional data.
b/ 081 includes 089.
c/ 091 includes 099.

Israel

ISIC Revision 4		Number of enterprises					Number of employees					Wages and salaries paid to employees				
			(number)					(thousands)					(millions of Israeli New Sheqalim)			
ISIC	Industry	Note	2010	2011	2012	2013	Note	2010	2011	2012	2013	Note	2010	2011	2012	2013
05	Mining of coal and lignite		...	-	-	-		...	...	...	...		...	-	-	...
06	Extraction of crude petroleum & natural gas		...	...	...	...		...	...	...	...		...	...	...	...
07	Mining of metal ores		...	...	-	-		...	...	...	...		...	...	-	...
08	Other mining and quarrying		...	102	101	...		...	3	3	...		...	845	1001	...
09	Mining support service activities		...	42	...	90		...	-	...	-		...	42	...	85
B	Mining and quarrying		...	148	176	191		...	4	4	4		...	925	1216	1055
35	Electricity, gas, steam & air conditioning		...	203	314	373		...	14	14	14		...	3880	4058	4101
D	Electricity, gas, steam & air conditioning		...	203	314	373		...	14	14	14		...	3880	4058	4101
36	Water collection, treatment and supply		...	267	267	277		...	4	4	4		...	694	747	794
37	Sewerage		...	144	147	...		...	1	1	...		...	130	147	...
38	Waste collection,treatment,disposal activities		...	359	365	390		...	6	6	7		...	529	573	632
39	Remediation activities		...	-	-	...		...	...	...	...		...	-	-	...
E	Water supply;sewerage,waste management		...	770	779	817		...	10	11	12		...	1354	1467	1571

Israel

ISIC Revision 4		Output (valuation not defined)					Value added (valuation not defined)				
			(millions of Israeli New Sheqalim)					(millions of Israeli New Sheqalim)			
ISIC	Industry	Note	2010	2011	2012	2013	Note	2010	2011	2012	2013
05	Mining of coal and lignite		...	...	...	...		...	...	...	...
06	Extraction of crude petroleum & natural gas		...	...	...	...		...	...	...	...
07	Mining of metal ores		...	...	...	...		...	...	...	...
08	Other mining and quarrying		...	...	...	...		...	...	...	...
09	Mining support service activities		...	...	...	...		...	...	...	...
B	Mining and quarrying		...	13207	14210	...		...	9772	7497	...
35	Electricity, gas, steam & air conditioning		...	29654a/	37489a/	...		...	10073a/	8689a/	...
D	Electricity, gas, steam & air conditioning		...	...	...	...		...	...	...	...
36	Water collection, treatment and supply		...	...a/	...a/	...		...	...a/	...a/	...
37	Sewerage		...	...a/	...a/	...		...	...a/	...a/	...
38	Waste collection,treatment,disposal activities		...	3975b/	4642b/	...		...	1196b/	1431b/	...
39	Remediation activities		...	...b/	...b/	...		...	...b/	...b/	...
E	Water supply;sewerage,waste management		...	...	...	...		...	...	...	...

a/ 35 includes 36 and 37.
b/ 38 includes 39.

Italy

ISIC Revision 4		Number of enterprises					Number of employees					Wages and salaries paid to employees				
			(number)					(thousands)					(millions of Euros)			
ISIC	Industry	Note	2010	2011	2012	2013	Note	2010	2011	2012	2013	Note	2010	2011	2012	2013
051	Mining of hard coal		...	...	...	...		...	...	...	...		...	...	...	...
052	Mining of lignite		-	-	-	-		-	-	-	-		-	-	-	-
061	Extraction of crude petroleum		3a/	3	4	5		12.1a/	0.3	0.3	0.4		740a/	23	24	29
062	Extraction of natural gas		...a/	5	8	7		...a/	13.0	12.6	12.0		...a/	742	722	793
071	Mining of iron ores		-	-	-	-		-	-	-	-		-	-	-	-
072	Mining of non-ferrous metal ores		...	...	...	...		...	...	...	...		...	...	...	...
081	Quarrying of stone, sand and clay		2243	2225	2119	2025		14.6	13.8	12.6	12.0		368	390	323	312
089	Mining and quarrying n.e.c.		238	223	268	249		2.3	2.2	2.7	2.5		66	69	81	73
091	Support activities for petroleum		34b/	...	44	42		1.6b/	...	1.7	1.8		79b/	...	94	100
099	Support activities for other mining, quarrying		...b/	-	-	-		...b/	-	-	-		...b/	-	-	-
B	Mining and quarrying		2522	2493	2451	2336		31.0	31.2	30.4	29.1		1268	1314	1262	1322
351	Electric power generation, transmission		3291	5758	8129	9336		56.6	56.0	58.5	57.6		2498	2502	2715	2680
352	Manufacture of gas		555	591	593	618		21.6	21.7	22.0	22.0		827	835	877	887
353	Steam and air conditioning supply		182	198	204	215		3.8	3.8	3.7	3.7		136	141	136	133
D	Electricity, gas, steam & air conditioning		4028	6547	8926	10169		82.0	81.5	84.3	83.3		3461	3478	3729	3700
360	Water collection, treatment and supply		865	849	858	873		28.2	29.2	28.6	30.0		948	1000	989	1043
370	Sewerage		1398	1396	1395	1428		8.9	8.6	8.9	8.4		276	267	271	256
381	Waste collection		2135	2166	2126	2123		90.7	90.0	91.1	89.4		2595	2559	2568	2581
382	Waste treatment and disposal		794	840	856	903		16.8	17.6	18.3	17.9		539	584	605	586
383	Materials recovery		3034	3105	3157	3199		20.2	20.7	22.0	22.4		478	501	554	572
390	Remediation activities		463	557	575	595		4.2	4.5	5.6	6.2		123	134	178	212
E	Water supply;sewerage,waste management		8689	8913	8967	9121		169.1	170.7	174.6	174.4		4959	5045	5164	5250

a/ 061 includes 062.
b/ 091 includes 099.

Italy

ISIC Revision 4		Output (valuation not defined)					Value added at factor values				
		Note	(millions of Euros)				Note	(millions of Euros)			
ISIC	Industry		2010	2011	2012	2013		2010	2011	2012	2013
051	Mining of hard coal		...	...	...	...		...	...	...	...
052	Mining of lignite		-	-	-	-		-	-	-	-
061	Extraction of crude petroleum		46805a/	1745	2164	1996		4354a/	1243	1510	1246
062	Extraction of natural gas		...a/	59416	66117	42607		...a/	2433	2058	1556
071	Mining of iron ores		-	-	-	-		-	-	-	-
072	Mining of non-ferrous metal ores		...	...	...	...		...	...	...	...
081	Quarrying of stone, sand and clay		2733	3065	2368	2196		981	1044	740	724
089	Mining and quarrying n.e.c.		571	571	691	614		162	166	182	167
091	Support activities for petroleum		372b/	...	493	488		183b/	...	203	212
099	Support activities for other mining, quarrying		...b/	-	-	-		...b/	-	-	-
B	Mining and quarrying		50502	65197	71870	47950		5683	5046	4707	3941
351	Electric power generation, transmission		98451	100229	88181	80780		19667	20014	23737	21502
352	Manufacture of gas		30272	39941	41633	14659		4120	4192	5372	4769
353	Steam and air conditioning supply		1956	1905	2062	2198		450	400	448	615
D	Electricity, gas, steam & air conditioning		130679	142076	131877	97637		24237	24606	29557	26886
360	Water collection, treatment and supply		6441	7059	7209	7543		2616	2868	2969	3305
370	Sewerage		1763	1625	1756	1585		651	682	644	605
381	Waste collection		10183	10833	10634	10224		4666	4629	4580	4553
382	Waste treatment and disposal		4175	4529	4902	5231		1639	1684	1802	1933
383	Materials recovery		8230	5848	8300	8091		1556	1452	1488	1406
390	Remediation activities		800	963	1091	1425		193	265	241	179
E	Water supply;sewerage,waste management		31591	30856	33891	34100		11320	11579	11507	11981

a/ 061 includes 062.
b/ 091 includes 099.

Japan

ISIC Revision 4		Number of establishments					Number of employees					Wages and salaries paid to employees				
			(number)					(number)					(millions of Japanese Yen)			
ISIC	Industry	Note	2009	2010	2011	2012	Note	2009	2010	2011	2012	Note	2009	2010	2011	2012
05	Mining of coal and lignite		...	...	...	...		...	...	...	...		...	...	...	...
06	Extraction of crude petroleum & natural gas		...	...	...	...		...	...	...	...		...	...	...	...
07	Mining of metal ores		...	...	...	...		...	...	...	...		...	...	...	...
08	Other mining and quarrying		80	75	...	...		1933	1886	...	...		5036	4903	...	4466
09	Mining support service activities		...	...	...	...		...	...	...	...		...	...	...	...
B	Mining and quarrying		...	...	...	...		...	...	...	...		...	...	...	4466
35	Electricity, gas, steam & air conditioning		181	177	...	...		1837	1807	...	...		877	839	...	1370
D	Electricity, gas, steam & air conditioning		181	177	...	...		1837	1807	...	...		877	839	...	1370
36	Water collection, treatment and supply		...	...	...	...		...	...	...	...		...	...	...	...
37	Sewerage		...	...	...	...		...	...	...	...		...	...	...	...
38	Waste collection,treatment,disposal activities		720	721	...	...		13929	13679	...	...		26027	24709	...	23897
39	Remediation activities		...	...	...	...		...	...	...	...		...	...	...	...
E	Water supply;sewerage,waste management		720	721	...	...		13929	13679	...	...		26027	24709	...	23897

Japan

ISIC Revision 4		Output (valuation not defined)					Value added (valuation not defined)				
			(millions of Japanese Yen)					(millions of Japanese Yen)			
ISIC	Industry	Note	2009	2010	2011	2012	Note	2009	2010	2011	2012
05	Mining of coal and lignite		...	...	...	...		...	...	...	...
06	Extraction of crude petroleum & natural gas		...	...	...	...		...	...	...	...
07	Mining of metal ores		...	...	...	...		...	...	...	...
08	Other mining and quarrying		50264	48739	...	42273		26433	24943	...	20500
09	Mining support service activities		...	...	...	...		...	...	...	...
B	Mining and quarrying		...	...	...	42273		...	...	...	20500
35	Electricity, gas, steam & air conditioning		21816	22356	...	24551		17349	16875	...	17685
D	Electricity, gas, steam & air conditioning		21816	22356	...	24551		17349	16875	...	17685
36	Water collection, treatment and supply		...	...	...	...		...	...	...	...
37	Sewerage		...	...	...	...		...	...	...	...
38	Waste collection,treatment,disposal activities		761388	878305	...	792942		201215	213912	...	190981
39	Remediation activities		...	...	...	...		...	...	...	...
E	Water supply;sewerage,waste management		761388	878305	...	792942		201215	213912	...	190981

Kazakhstan

ISIC Revision 4		Number of establishments					Number of employees					Wages and salaries paid to employees				
			(number)					(number)					(millions of Kazakhstan Tenge)			
ISIC	Industry	Note	2010	2011	2012	2013	Note	2010	2011	2012	2013	Note	2010	2011	2012	2013
05	Mining of coal and lignite		...	...	93	99		...	...	34035	33914		...	...	54.0	56.0
051	Mining of hard coal		...	...	81	87		...	...	31726	31599		...	...	51.8	53.4
052	Mining of lignite		...	...	12	12		...	...	2309	2315		...	...	2.3	2.6
06	Extraction of crude petroleum & natural gas		...	...	262	285		...	...	44349	45495		...	...	164.4	213.6
061	Extraction of crude petroleum		...	...	250	272		...	...	44349	45373		...	...	164.4	213.5
062	Extraction of natural gas		...	...	12	13		...	...	...	122		...	...	...	...
07	Mining of metal ores		...	...	304	353		...	...	85521	84386		...	...	135.0	150.5
071	Mining of iron ores		...	...	44	44		...	...	21674	22083		...	...	31.1	35.8
072	Mining of non-ferrous metal ores		...	...	260	309		...	...	63847	62303		...	...	103.9	114.7
08	Other mining and quarrying		...	...	1234	1355		...	...	16959	16434		...	...	16.0	16.4
081	Quarrying of stone, sand and clay		...	...	1067	1177		...	...	9999	9440		...	...	9.8	9.7
089	Mining and quarrying n.e.c.		...	...	167	178		...	...	6960	6994		...	...	6.2	6.7
09	Mining support service activities		...	...	597	692		...	...	35690	38528		...	...	93.2	114.5
091	Support activities for petroleum		...	...	523	580		...	...	32826	35296		...	...	87.4	107.6
099	Support activities for other mining, quarrying		...	...	74	112		...	...	2864	3232		...	...	5.8	6.9
B	Mining and quarrying		...	...	2490	2784		...	...	216554	218757		...	...	462.6	551.0
35	Electricity, gas, steam & air conditioning		...	...	1234	1268		...	...	115822	115880		...	...	127.5	145.0
351	Electric power generation, transmission		...	...	779	805		...	...	80389	80672		...	...	93.7	106.5
352	Manufacture of gas		...	...	159	163		...	...	8132	7527		...	...	11.3	12.6
353	Steam and air conditioning supply		...	...	296	300		...	...	27301	27681		...	...	22.6	26.0
D	Electricity, gas, steam & air conditioning		...	...	1234	1268		...	...	115822	115880		...	...	127.5	145.1
36	Water collection, treatment and supply		...	...	628	655		...	...	25145	27634		...	...	17.9	22.5
37	Sewerage		...	...	60	65		...	...	2701	2165		...	...	2.0	1.9
38	Waste collection,treatment,disposal activities		...	...	1306	1373		...	...	10912	11420		...	...	8.8	9.8
381	Waste collection		...	...	626	641		...	...	6602	7378		...	...	4.4	5.3
382	Waste treatment and disposal		...	...	216	244		...	...	2488	2246		...	...	2.8	2.9
383	Materials recovery		...	...	464	488		...	...	1822	1796		...	...	1.6	1.6
39	Remediation activities		...	...	71	84		...	...	214	262		...	...	...	...
E	Water supply;sewerage,waste management		...	...	2065	2177		...	...	38972	41481		...	...	292.5	312.5

Kazakhstan

ISIC Revision 4		Output (valuation not defined)					Value added (valuation not defined)				
			(millions of Kazakhstan Tenge)					(millions of Kazakhstan Tenge)			
ISIC	Industry	Note	2010	2011	2012	2013	Note	2010	2011	2012	2013
05	Mining of coal and lignite		...	286080	331353	335970		...	127209	136697	143962
051	Mining of hard coal		...	...	...	...		...	...	...	...
052	Mining of lignite		...	...	...	...		...	...	...	...
06	Extraction of crude petroleum & natural gas		...	8628312	8789779	9103777		...	4041674	4240751	4352073
061	Extraction of crude petroleum		...	...	...	...		...	...	...	...
062	Extraction of natural gas		...	...	...	...		...	...	...	...
07	Mining of metal ores		...	1302275	1467804	1692050		...	577947	607384	661839
071	Mining of iron ores		...	...	...	...		...	...	...	...
072	Mining of non-ferrous metal ores		...	...	...	...		...	...	...	...
08	Other mining and quarrying		...	131668	148650	131963		...	81127	84164	84749
081	Quarrying of stone, sand and clay		...	...	...	...		...	...	...	...
089	Mining and quarrying n.e.c.		...	...	...	...		...	...	...	...
09	Mining support service activities		...	408564	480589	553785		...	175296	219745	235071
091	Support activities for petroleum		...	...	...	...		...	...	...	...
099	Support activities for other mining, quarrying		...	...	...	...		...	...	...	...
B	Mining and quarrying		...	10756900	11218175	11817544		...	5003253	5288741	5477694
35	Electricity, gas, steam & air conditioning		...	...	...	...		...	...	...	...
351	Electric power generation, transmission		...	...	...	...		...	...	...	...
352	Manufacture of gas		...	...	...	...		...	...	...	...
353	Steam and air conditioning supply		...	...	...	...		...	...	...	...
D	Electricity, gas, steam & air conditioning		...	976669	1108819	1225279		...	478917	518617	580318
36	Water collection, treatment and supply		...	...	...	...		...	...	...	...
37	Sewerage		...	...	...	...		...	...	...	...
38	Waste collection,treatment,disposal activities		...	...	...	...		...	...	...	...
381	Waste collection		...	...	...	...		...	...	...	...
382	Waste treatment and disposal		...	...	...	...		...	...	...	...
383	Materials recovery		...	...	...	...		...	...	...	...
39	Remediation activities		...	...	...	...		...	...	...	...
E	Water supply;sewerage,waste management		...	228246	230634	220121		...	95086	99608	99498

Kuwait

ISIC Revision 3		Number of establishments					Number of employees					Wages and salaries paid to employees				
			(number)					(number)					(millions of Kuwaiti Dinars)			
ISIC	Industry	Note	2010	2011	2012	2013	Note	2010	2011	2012	2013	Note	2010	2011	2012	2013
101	Mining and agglomeration of hard coal		...	...	...	...		...	...	...	...		...	...	...	...
102	Mining and agglomeration of lignite		...	...	...	...		...	...	...	...		...	...	...	...
103	Extraction and agglomeration of peat		...	...	...	...		...	...	...	...		...	...	...	...
111	Extraction of crude oil and natural gas		3	3	3	3		10476	10853	11191	12221		390.5	533.9	614.2	618.5
112	Oil, gas service activities		4	4	4	4		4462	4207	4600	4812		30.2	34.5	35.9	39.3
120	Mining of uranium and thorium ores		...	...	...	...		...	...	...	...		...	...	...	...
131	Iron ores		...	...	...	...		...	...	...	...		...	...	...	...
132	Non-ferrous metal ores,excl.uranium,thorium		...	...	...	...		...	...	...	...		...	...	...	...
141	Quarrying of stone, sand and clay		...	...	...	...		...	...	...	...		...	...	...	...
142	Mining and quarrying n.e.c.		...	...	...	...		...	...	...	...		...	...	...	...
C	Total mining and quarrying	a/	7	7	7	7	a/	14938	15060	15791	17033	a/	420.6	568.4	650.1	657.7
401	Electricity production, collection, distribution		...	...	...	...		...	...	...	...		...	...	...	...
402	Gas production and distribution		...	...	...	...		...	...	...	...		...	...	...	...
403	Steam and hot water supply		...	...	...	...		...	...	...	...		...	...	...	...
410	Water collection, purification, distribution		...	...	...	...		...	...	...	...		...	...	...	...
E	Total electricity, gas and water supply		...	...	...	...		...	...	...	...		...	...	...	...

a/ Sum of available data.

Kuwait

ISIC Revision 3		Output at producers' prices					Value added at producers' prices				
			(millions of Kuwaiti Dinars)					(millions of Kuwaiti Dinars)			
ISIC	Industry	Note	2010	2011	2012	2013	Note	2010	2011	2012	2013
101	Mining and agglomeration of hard coal		...	...	...	...		...	...	...	...
102	Mining and agglomeration of lignite		...	...	...	...		...	...	...	...
103	Extraction and agglomeration of peat		...	...	...	...		...	...	...	...
111	Extraction of crude oil and natural gas		18788.4	27615.6	32697.7	32253.3		18406.5	26896.1	31764.8	31297.7
112	Oil, gas service activities		150.4	170.7	163.8	169.9		77.4	88.7	84.1	85.9
120	Mining of uranium and thorium ores		...	...	...	...		...	...	...	...
131	Iron ores		...	...	...	...		...	...	...	...
132	Non-ferrous metal ores,excl.uranium,thorium		...	...	...	...		...	...	...	...
141	Quarrying of stone, sand and clay		...	...	...	...		...	...	...	...
142	Mining and quarrying n.e.c.		...	...	...	...		...	...	...	...
C	Total mining and quarrying	a/	18938.7	27786.3	32861.5	32423.2	a/	18483.9	26984.8	31848.9	31383.6
401	Electricity production, collection, distribution		...	...	...	...		...	...	...	...
402	Gas production and distribution		...	...	...	...		...	...	...	...
403	Steam and hot water supply		...	...	...	...		...	...	...	...
410	Water collection, purification, distribution		...	...	...	...		...	...	...	...
E	Total electricity, gas and water supply		...	...	...	...		...	...	...	...

a/ Sum of available data.

Latvia

ISIC Revision 4		Number of establishments					Number of employees					Wages and salaries paid to employees				
			(number)					(number)					(thousands of Euros)			
ISIC	Industry	Note	2010	2011	2012	2013	Note	2010	2011	2012	2013	Note	2010	2011	2012	2013
051	Mining of hard coal		-	-	...	...		-	-	...	...		-	-	...	...
052	Mining of lignite		-	-	...	...		-	-	...	...		-	-	...	...
061	Extraction of crude petroleum		1	1	1	1		1	13	16	15		...	...	...	...
062	Extraction of natural gas		-	-	...	...		-	-	...	...		-	-	...	...
071	Mining of iron ores		-	-	...	...		-	-	...	...		-	-	...	...
072	Mining of non-ferrous metal ores		-	-	...	...		-	-	...	...		-	-	...	...
081	Quarrying of stone, sand and clay		140	137	162	168		643	724	784	895		3503	4560	3395	4290
089	Mining and quarrying n.e.c.		60	56	59	59		2010	1986	2197	2109		15741	17083	13089	14816
091	Support activities for petroleum		-	-	1	1		-	-	4	1		-	-	...	...
099	Support activities for other mining, quarrying		7	5	7	7		50	47	47	58		...	...	...	...
B	Mining and quarrying		208	199	230	236		5409	2770	3048	3078		19527	21949	16731	19364
351	Electric power generation, transmission		260	259	328	346		5242	5146	5287	5409		64872	71024	52652	54912
352	Manufacture of gas		3	3	6	10		1291	1310	1328	1313		...	...	...	...
353	Steam and air conditioning supply		119	121	119	124		4373	4181	4129	4383		...	...	...	...
D	Electricity, gas, steam & air conditioning		382	383	454	480		10906	10637	10744	11105		123706	131200	95200	99186
360	Water collection, treatment and supply		53	53	63	72		1759	1706	1701	1786		14459	15188	10798	11808
370	Sewerage		55	56	64	68		1436	1456	1473	1571		9361	9468	7011	7827
381	Waste collection		109	92	96	104		2662	2631	2894	3021		16827	16914	13457	14033
382	Waste treatment and disposal		21	22	30	30		290	310	362	458		3122	3391	2470	3014
383	Materials recovery		60	64	73	78		580	630	605	624		3044	3328	2662	2361
390	Remediation activities		13	11	15	19		68	70	81	86		374	309	228	257
E	Water supply;sewerage,waste management		312	298	342	372		6795	6803	7115	7547		47187	48597	36626	39300

Latvia

ISIC Revision 4		Output at producers' prices					Value added at factor values				
			(thousands of Euros)					(thousands of Euros)			
ISIC	Industry	Note	2010	2011	2012	2013	Note	2010	2011	2012	2013
051	Mining of hard coal		-	-	...	...		-	-	...	...
052	Mining of lignite		-	-	...	...		-	-	...	...
061	Extraction of crude petroleum		...	...	...	...		...	...	...	...
062	Extraction of natural gas		-	-	...	...		-	-	...	...
071	Mining of iron ores		-	-	...	...		-	-	...	...
072	Mining of non-ferrous metal ores		-	-	...	...		-	-	...	...
081	Quarrying of stone, sand and clay		34327	56714	44370	46565		13503	22348	15601	16215
089	Mining and quarrying n.e.c.		108252	110543	89035	100492		46088	45593	38658	39490
091	Support activities for petroleum		-	-	...	...		-	-	...	...
099	Support activities for other mining, quarrying		...	...	...	...		...	...	...	...
B	Mining and quarrying		144910	169463	134689	148696		59779	69105	54954	56552
351	Electric power generation, transmission		1177567	1258832	862587	883009		400971	389366	279270	277952
352	Manufacture of gas		...	...	...	...		...	...	...	...
353	Steam and air conditioning supply		...	...	...	...		...	...	...	...
D	Electricity, gas, steam & air conditioning		1592925	1689891	1178811	1225442		623253	609121	430393	459150
360	Water collection, treatment and supply		64808	66119	48367	47715		40119	40789	28126	30109
370	Sewerage		39910	39912	32400	34585		26633	26808	19209	21531
381	Waste collection		80779	84457	71779	71870		40240	34149	25947	31657
382	Waste treatment and disposal		19505	21246	17438	21441		11756	10639	8746	10122
383	Materials recovery		93874	103255	66991	39918		11847	11235	6471	4711
390	Remediation activities		2207	1762	1800	1800		730	710	772	702
E	Water supply;sewerage,waste management		301083	316751	238775	217329		131325	124332	89271	98832

Lithuania

ISIC Revision 4		Number of establishments					Number of employees					Wages and salaries paid to employees				
			(number)					(number)					(thousands of Euros)			
ISIC	Industry	Note	2010	2011	2012	2013	Note	2010	2011	2012	2013	Note	2010	2011	2012	2013
051	Mining of hard coal		-	-	-	-		-	-	-	-		-	-	-	-
052	Mining of lignite		-	-	-	-		-	-	-	-		-	-	-	-
061	Extraction of crude petroleum		4	4	4	4		252	261	279	269		3908	4141	4906	4903
062	Extraction of natural gas		-	-	-	-		-	-	-	-		-	-	-	-
071	Mining of iron ores		-	-	-	-		-	-	-	-		-	-	-	-
072	Mining of non-ferrous metal ores		-	-	-	-		-	-	-	-		-	-	-	-
081	Quarrying of stone, sand and clay		51	56	61	68		1008	1173	1158	1229		7315	9381	9425	11013
089	Mining and quarrying n.e.c.		24	27	25	25		1125	1133	1090	1090		7655	8450	7897	8757
091	Support activities for petroleum		-	-	-	-		-	-	-	-		-	-	-	-
099	Support activities for other mining, quarrying		-	-	-	-		-	-	-	-		-	-	-	-
B	Mining and quarrying		79	87	90	97		2385	2567	2527	2588		18878	21973	22229	24674
351	Electric power generation, transmission		165	201	400	1115		7884	5192	4670	5260		95784	56037	52197	54441
352	Manufacture of gas		11	8	8	14		1799	1761	1749	1980		23634	22837	23167	23990
353	Steam and air conditioning supply		77	85	80	85		6170	6332	6227	6240		52294	53524	54217	57380
D	Electricity, gas, steam & air conditioning		253	294	488	1214		15853	13285	12646	13480		171711	132399	129580	135810
360	Water collection, treatment and supply		75	73	76	75		5785	5697	5654	5620		41073	40633	41544	43183
370	Sewerage		40	41	44	47		576	600	570	665		3086	3311	3364	4717
381	Waste collection		92	105	115	117		2997	3232	3414	3459		16552	17536	18076	20003
382	Waste treatment and disposal		56	64	77	75		1346	1375	1574	1709		8744	9637	10665	12040
383	Materials recovery		73	72	83	78		1236	1483	1581	1543		7213	9078	10682	10753
390	Remediation activities		-	-	-	-		-	-	-	-		-	-	-	-
E	Water supply;sewerage,waste management		336	355	395	392		11940	12387	12793	12996		76667	80195	84330	90696

Lithuania

ISIC Revision 4		Output at producers' prices					Value added at factor values				
			(thousands of Euros)					(thousands of Euros)			
ISIC	Industry	Note	2010	2011	2012	2013	Note	2010	2011	2012	2013
051	Mining of hard coal		-	-	-	-		-	-	-	-
052	Mining of lignite		-	-	-	-		-	-	-	-
061	Extraction of crude petroleum		64705	77974	76026	63689		36086	51025	43554	33203
062	Extraction of natural gas		-	-	-	-		-	-	-	-
071	Mining of iron ores		-	-	-	-		-	-	-	-
072	Mining of non-ferrous metal ores		-	-	-	-		-	-	-	-
081	Quarrying of stone, sand and clay		50159	71449	66899	92546		19774	29409	28253	39024
089	Mining and quarrying n.e.c.		41201	50943	52704	60745		15383	20246	19218	22015
091	Support activities for petroleum		-	-	-	-		-	-	-	-
099	Support activities for other mining, quarrying		-	-	-	-		-	-	-	-
B	Mining and quarrying		156065	200366	195629	216980		71243	100680	91026	94242
351	Electric power generation, transmission		1766696	1204244	1129558	1269094		409806	292774	327250	378224
352	Manufacture of gas		152063	136082	130516	143090		107341	102854	95996	83664
353	Steam and air conditioning supply		658998	623327	660879	626587		146466	171270	170779	167302
D	Electricity, gas, steam & air conditioning		2577757	1963653	1920953	2038770		663614	566898	594025	629189
360	Water collection, treatment and supply		139072	138761	148829	157571		97821	101271	106415	116492
370	Sewerage		11718	12748	13301	29574		5673	5604	6206	10298
381	Waste collection		68091	103659	88755	95772		31346	34960	33829	36154
382	Waste treatment and disposal		50161	62501	70927	80089		23195	26881	29264	32961
383	Materials recovery		133762	159767	155018	133906		23356	29881	30728	26771
390	Remediation activities		-	-	-	-		-	-	-	-
E	Water supply;sewerage,waste management		402804	477436	476829	496911		181391	198597	206443	222676

Luxembourg

ISIC Revision 4		Number of enterprises					Number of employees					Wages and salaries paid to employees				
		Note	(number)				Note	(number)				Note	(millions of Euros)			
ISIC	Industry		2010	2011	2012	2013		2010	2011	2012	2013		2010	2011	2012	2013
051	Mining of hard coal		-	-	-	-		-	-	-	-		-	-	-	-
052	Mining of lignite		-	-	-	-		-	-	-	-		-	-	-	-
061	Extraction of crude petroleum		-	-	-	-		-	-	-	-		-	-	-	-
062	Extraction of natural gas		-	-	-	-		-	-	-	-		-	-	-	-
071	Mining of iron ores		-	-	-	-		-	-	-	-		-	-	-	-
072	Mining of non-ferrous metal ores		-	-	-	-		-	-	-	-		-	-	-	-
081	Quarrying of stone, sand and clay		10	10	10	10		284	285	275	266		13	13	12	12
089	Mining and quarrying n.e.c.		-	-	-	-		-	-	-	-		-	-	-	-
091	Support activities for petroleum		-	-	-	-		-	-	-	-		-	-	-	-
099	Support activities for other mining, quarrying		-	-	-	-		-	-	-	-		-	-	-	-
B	Mining and quarrying		10	10	10	10		284	285	275	266		13	13	12	12
351	Electric power generation, transmission		47	48	56	55		1196a/	1265a/	...	...		85a/	96a/	...	...
352	Manufacture of gas		6	6	4	5		...a/	...a/	...	...		...a/	...a/	...	...
353	Steam and air conditioning supply		14	14	14	14		...a/	...a/	155	168		...a/	...a/	7	8
D	Electricity, gas, steam & air conditioning		67	68	74	74		1196	1265	1341	1435		85	96	100	107
360	Water collection, treatment and supply		8	8	8	8		130	130	140	139		8	9	9	10
370	Sewerage		5	8	6	6		...	...	...	...		...	...	...	...
381	Waste collection		10	11	11	14		575	590	591	601		22	22	23	24
382	Waste treatment and disposal		11	11	13	13		101	117	127	129		6	6	7	7
383	Materials recovery		24	23	22	22		339	314	298	295		13	13	12	13
390	Remediation activities		3	3	4	4		...	...	...	...		...	...	...	...
E	Water supply;sewerage,waste management		61	64	64	67		1178	1185	1197	1210		50	52	52	55

a/ 351 includes 352 and 353.

Luxembourg

ISIC Revision 4		Output (valuation not defined)					Value added at factor values				
			(millions of Euros)					(millions of Euros)			
ISIC	Industry	Note	2010	2011	2012	2013	Note	2010	2011	2012	2013
051	Mining of hard coal		-	-	-	-		-	-	-	-
052	Mining of lignite		-	-	-	-		-	-	-	-
061	Extraction of crude petroleum		-	-	-	-		-	-	-	-
062	Extraction of natural gas		-	-	-	-		-	-	-	-
071	Mining of iron ores		-	-	-	-		-	-	-	-
072	Mining of non-ferrous metal ores		-	-	-	-		-	-	-	-
081	Quarrying of stone, sand and clay		69	74	75	75		33	36	32	32
089	Mining and quarrying n.e.c.		-	-	-	-		-	-	-	-
091	Support activities for petroleum		-	-	-	-		-	-	-	-
099	Support activities for other mining, quarrying		-	-	-	-		-	-	-	-
B	Mining and quarrying		69	74	75	75		33	36	32	32
351	Electric power generation, transmission		920a/	1021a/	...	...		260a/	276a/	...	...
352	Manufacture of gas		...a/	...a/	...	...		...a/	...a/	...	...
353	Steam and air conditioning supply		...a/	...a/	82	89		...a/	...a/	28	29
D	Electricity, gas, steam & air conditioning		920	1021	950	959		260	276	225	276
360	Water collection, treatment and supply		44	44	42	43		23	24	23	24
370	Sewerage		...	...	...	...		...	...	...	...
381	Waste collection		79	87	85	95		39	40	39	45
382	Waste treatment and disposal		26	74	21	21		12	12	11	10
383	Materials recovery		51	56	49	47		24	27	24	22
390	Remediation activities		...	...	...	...		...	...	...	...
E	Water supply;sewerage,waste management		205	268	205	215		100	107	101	103

a/ 351 includes 352 and 353.

Malaysia

ISIC Revision 4		Number of establishments					Number of employees					Wages and salaries paid to employees				
			(number)					(number)					(millions of Malaysian Ringgits)			
ISIC	Industry	Note	2010	2011	2012	2013	Note	2010	2011	2012	2013	Note	2010	2011	2012	2013
05	Mining of coal and lignite		7	...	...	...		635	...	...	...		20.9	...	...	...
051	Mining of hard coal		7	...	...	...		635	...	...	...		20.9	...	...	...
052	Mining of lignite		...	...	...	...		...	...	...	...		...	...	...	...
06	Extraction of crude petroleum & natural gas		16	16	16	...		12919	13304	15015	16355		2512.1	2811.6	3691.8	4157.0
061	Extraction of crude petroleum		10	10	10	...		10714	10928	12409	...		2175.0	2474.2	3244.2	...
062	Extraction of natural gas		6	6	6	...		2205	2376	2606	...		337.0	337.4	447.6	...
07	Mining of metal ores		60	...	...	...		4307	...	...	...		120.1	...	...	...
071	Mining of iron ores		24	...	...	...		1736	...	...	...		37.7	...	...	...
072	Mining of non-ferrous metal ores		36	...	...	...		2571	...	...	...		82.4	...	...	...
08	Other mining and quarrying		335	...	...	...		14284	...	...	...		344.7	...	...	...
081	Quarrying of stone, sand and clay		325	...	...	...		14069	...	...	...		341.0	...	...	...
089	Mining and quarrying n.e.c.		...	...	...	...		...	...	...	...		...	...	...	...
09	Mining support service activities		10	...	...	...		215	...	...	...		3.8	...	...	...
091	Support activities for petroleum		...	...	...	...		...	...	...	...		...	...	...	...
099	Support activities for other mining, quarrying		...	...	...	...		...	...	...	...		...	...	...	...
B	Mining and quarrying	a/	428	16	16	...	a/	32360	13304	15015	16355	a/	3001.6	2811.6	3691.8	4157.0
35	Electricity, gas, steam & air conditioning		...	...	...	...		...	...	41975	...		...	...	2450.0	...
351	Electric power generation, transmission		...	...	...	...		...	...	...	...		...	...	2299.0	...
352	Manufacture of gas		...	...	...	...		...	...	...	...		...	...	40.0	...
353	Steam and air conditioning supply		...	...	...	...		...	...	...	...		...	...	111.0	...
D	Electricity, gas, steam & air conditioning		...	...	...	...		...	...	41975a/	...		...	...	2450.0a/	...
36	Water collection, treatment and supply		...	...	...	...		...	...	...	...		...	...	...	...
37	Sewerage		...	...	...	...		...	...	...	...		...	...	...	...
38	Waste collection,treatment,disposal activities		...	...	...	...		...	...	...	...		...	...	...	...
381	Waste collection		...	...	...	...		...	...	...	...		...	...	...	...
382	Waste treatment and disposal		...	...	...	...		...	...	...	...		...	...	...	...
383	Materials recovery		...	...	...	...		...	...	...	...		...	...	...	...
39	Remediation activities		...	...	...	...		...	...	...	...		...	...	...	...
E	Water supply;sewerage,waste management		...	...	...	...		...	...	...	...		...	...	...	...

a/ Sum of available data.

Malaysia

ISIC Revision 4		Output (valuation not defined) a/					Value added (valuation not defined) b/				
			(millions of Malaysian Ringgits)					(millions of Malaysian Ringgits)			
ISIC	Industry	Note	2010	2011	2012	2013	Note	2010	2011	2012	2013
05	Mining of coal and lignite		323.4	...	...	...		52.0	...	...	...
051	Mining of hard coal		323.4	...	...	...		52.0	...	...	...
052	Mining of lignite		...	...	...	...		...	...	...	...
06	Extraction of crude petroleum & natural gas		98121.5	109193.0	109369.8	115096.0		87008.2	93066.8	94441.6	98308.0
061	Extraction of crude petroleum		81336.8	87432.7	81221.6	...		72921.0	74727.3	69366.2	...
062	Extraction of natural gas		16784.7	21760.3	28148.2	...		14087.2	18339.5	25075.5	...
07	Mining of metal ores		1204.7	...	...	...		547.6	...	...	...
071	Mining of iron ores		338.7	...	...	...		122.6	...	...	...
072	Mining of non-ferrous metal ores		866.0	...	...	...		425.0	...	...	...
08	Other mining and quarrying		2967.9	...	...	...		1089.4	...	...	...
081	Quarrying of stone, sand and clay		2941.2	...	...	...		1073.4	...	...	...
089	Mining and quarrying n.e.c.		...	...	...	...		...	...	...	...
09	Mining support service activities		26.6	...	...	...		16.0	...	...	...
091	Support activities for petroleum		...	...	...	...		...	...	...	...
099	Support activities for other mining, quarrying		...	...	...	...		...	...	...	...
B	Mining and quarrying	c/	102644.2	109193.0	109369.8	115096.0	c/	88713.2	93066.8	94441.6	98308.0
35	Electricity, gas, steam & air conditioning		...	...	38945.0	...		...	...	21132.0	...
351	Electric power generation, transmission		...	...	36248.0	...		...	...	20280.0	...
352	Manufacture of gas		...	...	2109.0	...		...	...	552.0	...
353	Steam and air conditioning supply		...	...	588.0	...		...	...	299.0	...
D	Electricity, gas, steam & air conditioning		...	...	38945.0c/	...		...	...	21132.0c/	...
36	Water collection, treatment and supply		...	...	...	...		...	...	...	...
37	Sewerage		...	...	...	...		...	...	...	...
38	Waste collection,treatment,disposal activities		...	...	...	...		...	...	...	...
381	Waste collection		...	...	...	...		...	...	...	...
382	Waste treatment and disposal		...	...	...	...		...	...	...	...
383	Materials recovery		...	...	...	...		...	...	...	...
39	Remediation activities		...	...	...	...		...	...	...	...
E	Water supply;sewerage,waste management		...	...	...	...		...	...	...	...

a/ Output is at basic prices.
b/ Value added is at basic prices.
c/ Sum of available data.

Mauritius

ISIC Revision 4		Number of establishments					Number of employees					Wages and salaries paid to employees				
			(number)					(number)					(millions of Mauritian Rupees)			
ISIC	Industry	Note	2009	2010	2011	2012	Note	2009	2010	2011	2012	Note	2009	2010	2011	2012
B	Mining and quarrying		27	25	27	25		1129	1116	1125	975		405.0	457.6	414.6	473.0
D	Electricity, gas, steam & air conditioning		7	7	7	7		2130	2287	2254	2373		888.0	1086.0	1324.0	1245.0
E	Water supply;sewerage,waste management		10	12	11	14		1709	1774	1721	1921		478.0	508.0	541.0	616.0

Mauritius

ISIC Revision 4		Output (valuation not defined) a/					Value added (valuation not defined) b/				
			(millions of Mauritian Rupees)					(millions of Mauritian Rupees)			
ISIC	Industry	Note	2009	2010	2011	2012	Note	2009	2010	2011	2012
B	Mining and quarrying		2909.6	3265.0	2929.5	2861.2		1040.2	1173.2	1040.4	999.7
D	Electricity, gas, steam & air conditioning		16517.0	16873.0	18420.0	19285.0		4786.0	4729.3	4491.4	4092.0
E	Water supply;sewerage,waste management		1447.0	1609.3	1653.5	2130.0		829.2	906.1	892.7	1193.0

a/ Output is reported at basic prices.
b/ Value added is reported at basic prices.

Mexico

ISIC Revision 4		Number of establishments					Number of persons engaged					Wages and salaries paid to employees				
			(number)					(number)					(millions of Mexican Pesos)			
ISIC	Industry	Note	2009	2010	2011	2012	Note	2009	2010	2011	2012	Note	2009	2010	2011	2012
051	Mining of hard coal		...	...	...	...		...	...	...	...		...	...	...	...
052	Mining of lignite		...	...	...	...		...	...	...	...		...	...	...	...
061	Extraction of crude petroleum		41a/	...	...	...		50273a/	...	...	...		25926a/	...	...	...
062	Extraction of natural gas		...a/	...	...	...		...a/	...	...	...		...a/	...	...	...
071	Mining of iron ores		...	...	...	...		...	...	...	...		...	...	...	...
072	Mining of non-ferrous metal ores		...	...	...	...		...	...	...	...		...	...	...	...
081	Quarrying of stone, sand and clay		...	...	...	...		...	...	...	...		...	...	...	...
089	Mining and quarrying n.e.c.		...	...	...	...		...	...	...	...		...	...	...	...
091	Support activities for petroleum		125b/	...	...	...		16124b/	...	...	...		940b/	...	...	...
099	Support activities for other mining, quarrying		...b/	...	...	...		...b/	...	...	...		...b/	...	...	...
B	Mining and quarrying		2957	...	...	...		142325	...	...	...		33185	12960	14040	15475
351	Electric power generation, transmission		65	...	...	...		124782	...	...	...		35521	...	...	...
352	Manufacture of gas		...	...	...	...		...	...	...	...		...	...	...	...
353	Steam and air conditioning supply		...	...	...	...		...	...	...	...		...	...	...	...
D	Electricity, gas, steam & air conditioning		...	...	...	...		...	...	...	...		...	...	...	...
360	Water collection, treatment and supply		...	...	...	...		...	...	...	...		...	...	...	...
370	Sewerage		...	...	...	...		...	...	...	...		...	...	...	...
381	Waste collection		...	...	...	...		...	...	...	...		...	...	...	...
382	Waste treatment and disposal		...	...	...	...		...	...	...	...		...	...	...	...
383	Materials recovery		...	...	...	...		...	...	...	...		...	...	...	...
390	Remediation activities		...	...	...	...		...	...	...	...		...	...	...	...
E	Water supply;sewerage,waste management		...	...	...	...		...	...	...	...		...	...	...	...

a/ 061 includes 062.
b/ 091 includes 099.

Mexico

ISIC Revision 4			Output (valuation not defined)					Value added (valuation not defined)			
			(millions of Mexican Pesos)					(millions of Mexican Pesos)			
ISIC	Industry	Note	2009	2010	2011	2012	Note	2009	2010	2011	2012
051	Mining of hard coal		...	...	...	...		...	...	...	...
052	Mining of lignite		...	...	...	...		...	...	...	...
061	Extraction of crude petroleum		1188512a/	...	...	...		1003189a/	...	...	...
062	Extraction of natural gas		...a/	...	...	...		...a/	...	...	...
071	Mining of iron ores		...	...	...	...		...	...	...	...
072	Mining of non-ferrous metal ores		...	...	...	...		...	...	...	...
081	Quarrying of stone, sand and clay		...	...	...	...		...	...	...	...
089	Mining and quarrying n.e.c.		...	...	...	...		...	...	...	...
091	Support activities for petroleum		24107b/	...	...	...		12474b/	...	...	...
099	Support activities for other mining, quarrying		...b/	...	...	...		...b/	...	...	...
B	Mining and quarrying		1293421	191467	264156	294938		1058627	135785	195738	216171
351	Electric power generation, transmission		446628	...	...	...		212770	...	...	...
352	Manufacture of gas		...	...	...	...		...	...	...	...
353	Steam and air conditioning supply		...	...	...	...		...	...	...	...
D	Electricity, gas, steam & air conditioning		...	...	...	...		...	...	...	...
360	Water collection, treatment and supply		...	...	...	...		...	...	...	...
370	Sewerage		...	...	...	...		...	...	...	...
381	Waste collection		...	...	...	...		...	...	...	...
382	Waste treatment and disposal		...	...	...	...		...	...	...	...
383	Materials recovery		...	...	...	...		...	...	...	...
390	Remediation activities		...	...	...	...		...	...	...	...
E	Water supply;sewerage,waste management		...	...	...	...		...	...	...	...

a/ 061 includes 062.
b/ 091 includes 099.

Mongolia

ISIC Revision 4		Number of establishments					Number of employees					Wages and salaries paid to employees				
			(number)					(number)					(millions of Mongolian Togrog)			
ISIC	Industry	Note	2010	2011	2012	2013	Note	2010	2011	2012	2013	Note	2010	2011	2012	2013
05	Mining of coal and lignite		...	...	...	...		5021	5549	7093	8847		46653	75556	120134	109017
06	Extraction of crude petroleum & natural gas		...	6	9	13		283	433	414	726		2755	7982	6952	8445
07	Mining of metal ores		...	191	187	174		13238	13420	12981	11881		142809	197957	239988	517099
08	Other mining and quarrying		...	115	135	115		2662	2814	2917	3334		18277	17120	25746	25672
09	Mining support service activities		...	406	329	318		2	5684	6667	10143		49635	292636	396258	118042
B	Mining and quarrying	a/	...	718	660	620	a/	21206	27900	30072	34931	a/	260129	591251	789078	778275
35	Electricity, gas, steam & air conditioning		...	173	187	178		14123	11546	13157	13495		86102	114424	138349	168927
D	Electricity, gas, steam & air conditioning	a/	...	173	187	178	a/	14123	11546	13157	13495	a/	86102	114424	138349	168927
36	Water collection, treatment and supply		...	46	53	83		3964	3669	5333	4925		16898	24912	33972	34307
37	Sewerage		...	57	75	34		3198	1912	2126	863		111	542	1182	1468
38	Waste collection,treatment,disposal activities		...	26	25	49		41	114	100	1243		3399	7236	10213	15799
39	Remediation activities		...	22	16	7		40	374	339	12		484	31	90	63
E	Water supply;sewerage,waste management	a/	...	151	169	173	a/	7243	6069	7898	7043	a/	20892	32721	45457	51637

a/ Sum of available data.

Mongolia

ISIC Revision 4		Output (valuation not defined) a/					Value added (valuation not defined) b/				
			(millions of Mongolian Togrog)					(millions of Mongolian Togrog)			
ISIC	Industry	Note	2010	2011	2012	2013	Note	2010	2011	2012	2013
05	Mining of coal and lignite		931511	2145640	1722628	1461125		547372	954644	665373	593788
06	Extraction of crude petroleum & natural gas		164391	273432	430656	770940		55848	106721	163337	305561
07	Mining of metal ores		1694958	2172324	2255636	3273706		1037542	1006904	1274962	1740617
08	Other mining and quarrying		101019	132209	154078	113897		41257	47914	45209	36107
09	Mining support service activities		1228233	962702	1224027	480088		420138	420056	593869	173342
B	Mining and quarrying	c/	4120112	5686307	5787025	6099756	c/	2102157	2536239	2742750	2849415
35	Electricity, gas, steam & air conditioning		636759	677377	824143	904495		190212	212394	233859	270514
D	Electricity, gas, steam & air conditioning	c/	636759	677377	824143	904495	c/	190212	212394	233859	270514
36	Water collection, treatment and supply		87091	103157	135844	146020		25361	33941	52066	45796
37	Sewerage		317	4393	10171	5643		162	751	1702	1870
38	Waste collection,treatment,disposal activities		18582	31914	50470	73944		5188	10017	18302	29607
39	Remediation activities		1293	1392	1626	1068		500	286	460	262
E	Water supply;sewerage,waste management	c/	107283	140856	198111	226675	c/	31211	44995	72530	77535

a/ Output is at basic prices.
b/ Value added is at basic prices.
c/ Sum of available data.

Netherlands

ISIC Revision 4		Number of enterprises					Number of employees					Wages and salaries paid to employees			
			(number)					(number)					(millions of Euros)		
ISIC	Industry	Note	2010	2011	2012	2013	Note	2010	2011	2012	2013	Note	2010	2011	2012	2013
05	Mining of coal and lignite		-	-	-	-		-	-	-	-		-	-	-	-
051	Mining of hard coal		-	-	-	-		-	-	-	-		-	-	-	-
052	Mining of lignite		-	-	-	-		-	-	-	-		-	-	-	-
06	Extraction of crude petroleum & natural gas		48	43	33	45		3160	3223	3376	3615		246	298	310	327
061	Extraction of crude petroleum		38	33	26	34		515	605	625	577	a/	246	298	310	327
062	Extraction of natural gas		10	10	7	11		2645	2618	2751	3038	a/	...	...	...	...
07	Mining of metal ores		-	-	-	-		-	-	-	-		-	-	-	-
071	Mining of iron ores		-	-	-	-		-	-	-	-		-	-	-	-
072	Mining of non-ferrous metal ores		-	-	-	-		-	-	-	-		-	-	-	-
08	Other mining and quarrying		147	150	144	162		1829	2335	2092	2122		80	97	83	81
081	Quarrying of stone, sand and clay		117	120	114	128		852	861	781	786		43	48	38	36
089	Mining and quarrying n.e.c.		30	30	30	34		977	1474	1311	1336		37	49	45	45
09	Mining support service activities		116	116	144	210		2663	2821	3125	4528		194	214	268	365
091	Support activities for petroleum		116	115	139	201	b/	2663	2821	3125	4528		...	214b/	...	365b/
099	Support activities for other mining, quarrying		-	1	5	9	b/	...	...	...	...		-	...b/	-	...b/
B	Mining and quarrying		311	309	321	417		7651	8379	8593	10265		520	610	661	773
35	Electricity, gas, steam & air conditioning		678	702	765	950		22206	23339	25799	26241		1342	1415	1419	1530
351	Electric power generation, transmission		673	697	757	940	c/	22206	23339	25799	26241		1342c/	1415c/	1415	1529
352	Manufacture of gas		4	4	6	6	c/	...	...	...	...		...c/	...c/	...	...
353	Steam and air conditioning supply		1	1	2	4	c/	...	...	...	...		...c/	...c/	...	...
D	Electricity, gas, steam & air conditioning		678	702	765	950		22206	23339	25799	26241		1342	1415	1419	1530
36	Water collection, treatment and supply		30	29	24	29		5283	5237	5291	5329		259	267	272	282
37	Sewerage		121	130	139	138		4632	4680	4483	3867		156	166	165	171
38	Waste collection,treatment,disposal activities		782	788	806	892		25850	26043	26321	21535		825	904	849	884
381	Waste collection		283	292	288	269		16519	16032	15840	10967		430	436	399	402
382	Waste treatment and disposal		223	208	217	247		5888	6119	6211	6258		249	285	271	303
383	Materials recovery		276	288	301	376		3443	3891	4270	4310		146	183	180	179
39	Remediation activities		218	222	250	386		1700	1755	1604	1613		64	78	74	67
E	Water supply;sewerage,waste management		1151	1169	1219	1445		37465	37714	37699	32343		1304	1415	1361	1405

a/ 061 includes 062.
b/ 091 includes 099.
c/ 351 includes 352 and 353.

Netherlands

ISIC Revision 4		Output (valuation not defined)					Value added at factor values				
		Note	(millions of Euros)				Note	(millions of Euros)			
ISIC	Industry		2010	2011	2012	2013		2010	2011	2012	2013
05	Mining of coal and lignite		-	-	-	-		-	-	-	-
051	Mining of hard coal		-	-	-	-		-	-	-	-
052	Mining of lignite		-	-	-	-		-	-	-	-
06	Extraction of crude petroleum & natural gas		16151	17706	21194	21855		8603	8966	9993	9547
061	Extraction of crude petroleum	a/	16151	17706	21194	21855	a/	8603	8966	9993	9547
062	Extraction of natural gas	a/	...	...	...	...	a/	...	...	...	...
07	Mining of metal ores		-	-	-	-		-	-	-	-
071	Mining of iron ores		-	-	-	-		-	-	-	-
072	Mining of non-ferrous metal ores		-	-	-	-		-	-	-	-
08	Other mining and quarrying		850	961	934	858		284	286	233	207
081	Quarrying of stone, sand and clay		482	457	431	370		189	157	114	110
089	Mining and quarrying n.e.c.		368	504	503	488		95	129	119	97
09	Mining support service activities		2246	2409	2806	3449		843	838	1098	1295
091	Support activities for petroleum		...	2409b/	...	3449b/		...	838b/	...	1295b/
099	Support activities for other mining, quarrying		-	...b/	-	...b/		-	...b/	-	...b/
B	Mining and quarrying		19246	21077	24933	26162		9729	10090	11325	11048
35	Electricity, gas, steam & air conditioning		38970	38527	27396	25572		6947	7654	7357	8170
351	Electric power generation, transmission		38970c/	38527c/	27394	25563		6947c/	7654c/	7366	8168
352	Manufacture of gas		...c/	...c/	...	...		...c/	...c/	...	...
353	Steam and air conditioning supply		...c/	...c/	...	...		...c/	...c/	...	...
D	Electricity, gas, steam & air conditioning		38970	38527	27396	25572		6947	7654	7357	8170
36	Water collection, treatment and supply		1596	1597	1477	1500		906	942	951	977
37	Sewerage		638	658	649	663		250	260	258	267
38	Waste collection,treatment,disposal activities		6095	6769	6742	6700		2140	2254	2131	2093
381	Waste collection		2336	2319	2158	2093		826	838	754	729
382	Waste treatment and disposal		1843	2029	1848	2069		891	927	901	968
383	Materials recovery		1915	2421	2736	2539		423	490	477	395
39	Remediation activities		372	430	409	429		115	120	143	142
E	Water supply;sewerage,waste management		8701	9455	9277	9292		3411	3576	3482	3478

a/ 061 includes 062.
b/ 091 includes 099.
c/ 351 includes 352 and 353.

New Zealand

ISIC Revision 4		Number of enterprises					Number of persons engaged					Wages and salaries paid to employees				
			(number)					(number)					(millions of New Zealand Dollars)			
ISIC	Industry	Note	2010	2011	2012	2013	Note	2010	2011	2012	2013	Note	2010	2011	2012	2013
051	Mining of hard coal		20	19	19	20		1910	1940	1600	1250		...	...	...	...
052	Mining of lignite		...	...	...	...		...	...	...	...		...	...	...	...
061	Extraction of crude petroleum		8	8	7	8		520	610	990	1190		...	...	...	...
062	Extraction of natural gas		...	...	...	...		...	...	...	...		...	...	...	...
071	Mining of iron ores		1	1	1	1		100	110	150	160		364a/	476a/	481a/	...
072	Mining of non-ferrous metal ores		...	...	...	...		...	...	...	...		...a/	...a/	...a/	...
081	Quarrying of stone, sand and clay		125	116	120	115		1080	1030	1030	1040		90b/	93b/	109b/	...
089	Mining and quarrying n.e.c.		...	...	...	...		...	...	...	...		...b/	...b/	...b/	...
091	Support activities for petroleum		53	64	64	63		950	1070	1180	1170		118c/	119c/	134c/	...
099	Support activities for other mining, quarrying		10	11	10	9		85	140	80	90		...c/	...c/	...c/	...
B	Mining and quarrying		...	...	...	...		...	...	...	...		572	688	724	...
351	Electric power generation, transmission		53	54	57	62		8670	9000	9180	9740		689d/	709d/	787d/	...
352	Manufacture of gas		4	4	3	2		190	200	60	18		...d/	...d/	...d/	...
353	Steam and air conditioning supply		...	...	...	...		...	...	...	...		...d/	...d/	...d/	...
D	Electricity, gas, steam & air conditioning		...	...	...	...		...	...	...	...		689	709	787	...
360	Water collection, treatment and supply		20	25	29	31		780	830	920	970		...	...	...	...
370	Sewerage		4	3	4	10		18	12	12	35		...	...	...	...
381	Waste collection		...	...	...	...		...	...	...	...		113e/	118e/	129e/	...
382	Waste treatment and disposal		...	...	...	...		...	...	...	...		...e/	...e/	...e/	...
383	Materials recovery		...	...	...	...		...	...	...	...		...e/	...e/	...e/	...
390	Remediation activities		77	76	83	90		1780	1740	1940	1850		76	95	97	...
E	Water supply;sewerage,waste management		...	...	...	...		...	...	...	...		228	257	275	...

a/ 071 includes 072.
b/ 081 includes 089.
c/ 091 includes 099.
d/ 351 includes 352 and 353.
e/ 381 includes 382 and 383.

New Zealand

ISIC Revision 4		Output (valuation not defined)					Value added (valuation not defined)				
			(millions of New Zealand Dollars)					(millions of New Zealand Dollars)			
ISIC	Industry	Note	2010	2011	2012	2013	Note	2010	2011	2012	2013
051	Mining of hard coal		...	...	...	...		...	...	...	...
052	Mining of lignite		...	...	...	...		...	...	...	...
061	Extraction of crude petroleum		...	...	...	...		...	...	...	...
062	Extraction of natural gas		...	...	...	...		...	...	...	...
071	Mining of iron ores		...	...	...	...		...	...	...	...
072	Mining of non-ferrous metal ores		...	...	...	...		...	...	...	...
081	Quarrying of stone, sand and clay		571a/	570a/	643a/	...		227a/	210a/	285a/	...
089	Mining and quarrying n.e.c.		...a/	...a/	...a/	...		...a/	...a/	...a/	...
091	Support activities for petroleum		683b/	671b/	732b/	...		320b/	274b/	225b/	...
099	Support activities for other mining, quarrying		...b/	...b/	...b/	...		...b/	...b/	...b/	...
B	Mining and quarrying		7605	8195	7778	...		3767	3679	3410	...
351	Electric power generation, transmission		13891c/	15834c/	16214c/	...		4845c/	5103c/	5423c/	...
352	Manufacture of gas		...c/	...c/	...c/	...		...c/	...c/	...c/	...
353	Steam and air conditioning supply		...c/	...c/	...c/	...		...c/	...c/	...c/	...
D	Electricity, gas, steam & air conditioning		13891	15834	16214	...		4845	5103	5423	...
360	Water collection, treatment and supply		...	...	...	...		...	...	...	...
370	Sewerage		...	...	...	...		...	...	...	...
381	Waste collection		578d/	604d/	678d/	...		295d/	289d/	311d/	...
382	Waste treatment and disposal		...d/	...d/	...d/	...		...d/	...d/	...d/	...
383	Materials recovery		...d/	...d/	...d/	...		...d/	...d/	...d/	...
390	Remediation activities		446	503	525	...		...	...	...	...
E	Water supply;sewerage,waste management		1487	1608	1748	...		838	822	906	...

a/ 081 includes 089.
b/ 091 includes 099.
c/ 351 includes 352 and 353.
d/ 381 includes 382 and 383.

Norway

ISIC Revision 4		Number of enterprises					Number of employees					Wages and salaries paid to employees				
			(number)					(number)					(millions of Norwegian Kroner)			
ISIC	Industry	Note	2010	2011	2012	2013	Note	2010	2011	2012	2013	Note	2010	2011	2012	2013
051	Mining of hard coal		1	1	1	1		361	358	402	378		262	260	280	279
052	Mining of lignite		-	-	-	-		-	-	-	-		-	-	-	-
061	Extraction of crude petroleum		61	68	68	71		21971	25610	26452	27297		...	23890a/	28120a/	28477
062	Extraction of natural gas		5	6	5	4		1264	1256	1350	1423		...	...a/	...a/	1878
071	Mining of iron ores		3	4	6	6		472	557	612	644		...	461b/	516b/	572b/
072	Mining of non-ferrous metal ores		3	3	3	4		286	293	298	307		...	...b/	...b/	...b/
081	Quarrying of stone, sand and clay		575	610	610	623		2723	2860	2996	2978		1104	1228	1342	1365
089	Mining and quarrying n.e.c.		36	35	38	30		525	536	539	406		251	241	256	212
091	Support activities for petroleum		297	323	332	332		26270	31507	33054	36355		17502	21006	22160	28804
099	Support activities for other mining, quarrying		23	20	17	17		30	34	30	18		10	16	13	10
B	Mining and quarrying		1004	1070	1080	1088		53902	63011	65733	69806		40886	47103	52688	61597
351	Electric power generation, transmission		876	914	951	962		14671	14553	14828	15143		6492	6885	8576	8170
352	Manufacture of gas		21	21	21	26		37	38	55	65		23	26	43	44
353	Steam and air conditioning supply		115	118	124	136		381	467	467	535		222	253	281	313
D	Electricity, gas, steam & air conditioning		1012	1053	1096	1124		15089	15058	15350	15743		6737	7164	8900	8526
360	Water collection, treatment and supply		441	448	462	501		319	325	320	343		138	83	84	105
370	Sewerage		147	145	157	165		636	672	678	684		272	300	315	319
381	Waste collection		238	230	230	237		4246	4322	4220	4263		1585	1730	1733	1796
382	Waste treatment and disposal		63	68	69	72		838	730	771	846		391	362	404	455
383	Materials recovery		141	145	131	121		1795	1795	1787	1756		825	831	851	842
390	Remediation activities		36	34	35	36		67	90	80	53		43	67	45	22
E	Water supply;sewerage,waste management		1066	1070	1084	1132		7901	7934	7856	7945		3255	3374	3430	3540

a/ 061 includes 062.
b/ 071 includes 072.

Norway

ISIC Revision 4		Output (valuation not defined)					Value added at factor values				
			(millions of Norwegian Kroner)					(millions of Norwegian Kroner)			
ISIC	Industry	Note	2010	2011	2012	2013	Note	2010	2011	2012	2013
051	Mining of hard coal		1642	1310	851	1171		864	549	185	421
052	Mining of lignite		-	-	-	-		-	-	-	-
061	Extraction of crude petroleum		...	1175502a/	1253985a/	1103261		...	562062a/	607306a/	513581
062	Extraction of natural gas		...	...a/	...a/	48877		...	...a/	...a/	27227
071	Mining of iron ores		...	2612b/	2962b/	2954b/		...	1096b/	1356b/	1180b/
072	Mining of non-ferrous metal ores		...	...b/	...b/	...b/		...	...b/	...b/	...b/
081	Quarrying of stone, sand and clay		6674	7483	8208	8390		2112	2502	2744	2893
089	Mining and quarrying n.e.c.		1651	1803	1589	1476		711	787	658	591
091	Support activities for petroleum		89638	95228	111360	132190		33492	36326	39536	48707
099	Support activities for other mining, quarrying		49	66	47	35		10	26	29	20
B	Mining and quarrying		1051342	1284006	1379003	1298354		480418	603347	651814	594618
351	Electric power generation, transmission		137056	110566	106363	118364		54841	39203	45240	51484
352	Manufacture of gas		558	556	632	728		143	199	14	-504
353	Steam and air conditioning supply		3171	3032	3104	3688		1078	1027	1114	1395
D	Electricity, gas, steam & air conditioning		140785	114155	110100	122781		56063	40429	46368	52375
360	Water collection, treatment and supply		388	418	391	399		182	181	168	161
370	Sewerage		983	1093	1118	1165		501	563	573	585
381	Waste collection		8983	9810	9365	9767		2968	3246	3240	3167
382	Waste treatment and disposal		2694	2317	2516	2709		983	886	980	1077
383	Materials recovery		7185	7941	8322	7879		1906	1990	1949	1740
390	Remediation activities		323	396	317	117		32	-1	163	59
E	Water supply;sewerage,waste management		20557	21974	22029	22035		6572	6864	7072	6789

a/ 061 includes 062.
b/ 071 includes 072.

Panama

ISIC Revision 4		Number of establishments					Number of employees					Wages and salaries paid to employees				
			(number)					(number)					(millions of Panamanian Balboas)			
ISIC	Industry	Note	2010	2011	2012	2013	Note	2010	2011	2012	2013	Note	2010	2011	2012	2013
05	Mining of coal and lignite		...	...	...	...		...	...	...	...		...	...	...	...
051	Mining of hard coal		...	...	...	...		...	...	...	...		...	...	...	...
052	Mining of lignite		...	...	...	...		...	...	...	...		...	...	...	...
06	Extraction of crude petroleum & natural gas		...	...	...	...		...	...	...	...		...	...	...	...
061	Extraction of crude petroleum		...	...	...	...		...	...	...	...		...	...	...	...
062	Extraction of natural gas		...	...	...	...		...	...	...	...		...	...	...	...
07	Mining of metal ores		...	...	...	...		...	...	...	...		...	...	...	...
071	Mining of iron ores		...	...	...	...		...	...	...	...		...	...	...	...
072	Mining of non-ferrous metal ores		...	...	...	...		...	...	...	...		...	...	...	...
08	Other mining and quarrying		...	25	27	29		...	723	757	1420		...	5.2	8.3	14.5
081	Quarrying of stone, sand and clay		...	25	23	24		...	723	660	667		...	5.2	7.0	7.5
089	Mining and quarrying n.e.c.		...	...	4	5		...	...	97	753		...	...	1.3	7.0
09	Mining support service activities		...	...	...	...		...	...	...	...		...	...	...	...
091	Support activities for petroleum		...	...	...	...		...	...	...	...		...	...	...	...
099	Support activities for other mining, quarrying		...	...	...	...		...	...	...	...		...	...	...	...
B	Mining and quarrying		...	25a/	27a/	29a/		...	723a/	757a/	1420a/		...	5.2a/	8.3a/	14.5a/
35	Electricity, gas, steam & air conditioning		...	23	...	...		...	2613	...	...		...	46.6	...	...
351	Electric power generation, transmission		...	14	...	...		...	2108	...	...		...	42.9	...	...
352	Manufacture of gas		...	...	...	...		...	...	...	...		...	...	...	...
353	Steam and air conditioning supply		...	9	...	...		...	505	...	...		...	3.7	...	...
D	Electricity, gas, steam & air conditioning		...	23a/	...	...		...	2613a/	...	...		...	46.6a/	...	...
36	Water collection, treatment and supply		...	...	...	...		...	...	...	...		...	...	...	...
37	Sewerage		...	...	...	...		...	...	...	...		...	...	...	...
38	Waste collection,treatment,disposal activities		...	12	...	...		...	778	...	...		...	5.7	...	...
381	Waste collection		...	9	...	...		...	579	...	...		...	4.1	...	...
382	Waste treatment and disposal		...	3	...	...		...	199	...	...		...	1.6	...	...
383	Materials recovery		...	...	...	...		...	...	...	...		...	...	...	...
39	Remediation activities		...	...	...	...		...	...	...	...		...	...	...	...
E	Water supply;sewerage,waste management		...	12a/	...	...		...	778a/	...	...		...	5.7a/	...	...

a/ Sum of available data.

Panama

ISIC Revision 4		Output at producers' prices					Value added				
			(millions of Panamanian Balboas)					(millions of Panamanian Balboas)			
ISIC	Industry	Note	2010	2011	2012	2013	Note	2010	2011	2012	2013
05	Mining of coal and lignite		...	...	...	...		...	...	...	...
051	Mining of hard coal		...	...	...	...		...	...	...	...
052	Mining of lignite		...	...	...	...		...	...	...	...
06	Extraction of crude petroleum & natural gas		...	...	...	...		...	...	...	...
061	Extraction of crude petroleum		...	...	...	...		...	...	...	...
062	Extraction of natural gas		...	...	...	...		...	...	...	...
07	Mining of metal ores		...	...	...	...		...	...	...	...
071	Mining of iron ores		...	...	...	...		...	...	...	...
072	Mining of non-ferrous metal ores		...	...	...	...		...	...	...	...
08	Other mining and quarrying		...	54.7	60.8	52.2		...	...	...	...
081	Quarrying of stone, sand and clay		...	54.7	60.1	51.4		...	...	...	...
089	Mining and quarrying n.e.c.		...	...	0.7	0.8		...	...	...	...
09	Mining support service activities		...	...	...	...		...	...	...	...
091	Support activities for petroleum		...	...	...	...		...	...	...	...
099	Support activities for other mining, quarrying		...	...	...	...		...	...	...	...
B	Mining and quarrying		...	54.7a/	60.8a/	52.2a/		...	...	...	...
35	Electricity, gas, steam & air conditioning		...	2288.8	...	...		...	...	...	...
351	Electric power generation, transmission		...	2263.2	...	...		...	...	...	...
352	Manufacture of gas		...	...	...	...		...	...	...	...
353	Steam and air conditioning supply		...	25.6	...	...		...	...	...	...
D	Electricity, gas, steam & air conditioning		...	2288.8a/	...	...		...	...	...	...
36	Water collection, treatment and supply		...	...	...	...		...	...	...	...
37	Sewerage		...	...	...	...		...	...	...	...
38	Waste collection,treatment,disposal activities		...	32.8	...	...		...	...	...	...
381	Waste collection		...	23.3	...	...		...	...	...	...
382	Waste treatment and disposal		...	9.5	...	...		...	...	...	...
383	Materials recovery		...	...	...	...		...	...	...	...
39	Remediation activities		...	...	...	...		...	...	...	...
E	Water supply;sewerage,waste management		...	32.8a/	...	...		...	...	...	...

a/ Sum of available data.

Paraguay

ISIC Revision 4		Number of establishments					Number of persons engaged					Wages and salaries paid to employees				
			(number)					(number)					(millions of Paraguayan Guaranies)			
ISIC	Industry	Note	2010	2011	2012	2013	Note	2010	2011	2012	2013	Note	2010	2011	2012	2013
05	Mining of coal and lignite		...	...	...	...		...	...	...	...		...	...	...	...
051	Mining of hard coal		...	...	...	...		...	...	...	...		...	...	...	...
052	Mining of lignite		...	...	...	...		...	...	...	...		...	...	...	...
06	Extraction of crude petroleum & natural gas		...	...	...	...		...	...	...	...		...	...	...	...
061	Extraction of crude petroleum		...	...	...	...		...	...	...	...		...	...	...	...
062	Extraction of natural gas		...	...	...	...		...	...	...	...		...	...	...	...
07	Mining of metal ores		...	...	...	...		...	...	...	...		...	...	...	...
071	Mining of iron ores		...	...	...	...		...	...	...	...		...	...	...	...
072	Mining of non-ferrous metal ores		...	...	...	...		...	...	...	...		...	...	...	...
08	Other mining and quarrying		80	...	...	...		998	...	...	...		26027	...	...	...
081	Quarrying of stone, sand and clay		80	...	...	...		998	...	...	...		26027	...	...	...
089	Mining and quarrying n.e.c.		...	...	...	...		...	...	...	...		...	...	...	...
09	Mining support service activities		16	...	...	...		43	...	...	...		439	...	...	...
091	Support activities for petroleum		...	...	...	...		...	...	...	...		...	...	...	...
099	Support activities for other mining, quarrying		16	...	...	...		43	...	...	...		439	...	...	...
B	Mining and quarrying		101	...	...	...		1123	...	...	...		28999	...	...	...
35	Electricity, gas, steam & air conditioning		7	...	...	...		8033	...	...	...		1797056	...	...	...
351	Electric power generation, transmission		7	...	...	...		8033	...	...	...		1797056	...	...	...
352	Manufacture of gas		...	...	...	...		...	...	...	...		...	...	...	...
353	Steam and air conditioning supply		...	...	...	...		...	...	...	...		...	...	...	...
D	Electricity, gas, steam & air conditioning		7a/	...	...	...		8033a/	...	...	...		1797056a/	...	...	...
36	Water collection, treatment and supply		516	...	...	...		3135	...	...	...		94448	...	...	...
37	Sewerage		8	...	...	...		49	...	...	...		440	...	...	...
38	Waste collection,treatment,disposal activities		67	...	...	...		833	...	...	...		11766	...	...	...
381	Waste collection		41	...	...	...		420	...	...	...		3866	...	...	...
382	Waste treatment and disposal		5	...	...	...		167	...	...	...		4488	...	...	...
383	Materials recovery		21	...	...	...		246	...	...	...		3413	...	...	...
39	Remediation activities		...	...	...	...		...	...	...	...		...	...	...	...
E	Water supply;sewerage,waste management		591a/	...	...	...		4017a/	...	...	...		106654a/	...	...	...

a/ Sum of available data.

Paraguay

ISIC Revision 4		Output at producers' prices					Value added at producers' prices				
		Note	(millions of Paraguayan Guaranies)				Note	(millions of Paraguayan Guaranies)			
ISIC	Industry		2010	2011	2012	2013		2010	2011	2012	2013
05	Mining of coal and lignite		...	...	...	...		...	...	...	...
051	Mining of hard coal		...	...	...	...		...	...	...	...
052	Mining of lignite		...	...	...	...		...	...	...	...
06	Extraction of crude petroleum & natural gas		...	...	...	...		...	...	...	...
061	Extraction of crude petroleum		...	...	...	...		...	...	...	...
062	Extraction of natural gas		...	...	...	...		...	...	...	...
07	Mining of metal ores		...	...	...	...		...	...	...	...
071	Mining of iron ores		...	...	...	...		...	...	...	...
072	Mining of non-ferrous metal ores		...	...	...	...		...	...	...	...
08	Other mining and quarrying		227706	...	...	...		103723	...	...	...
081	Quarrying of stone, sand and clay		227706	...	...	...		103723	...	...	...
089	Mining and quarrying n.e.c.		...	...	...	...		...	...	...	...
09	Mining support service activities		9437	...	...	...		5061	...	...	...
091	Support activities for petroleum		...	...	...	...		...	...	...	...
099	Support activities for other mining, quarrying		9437	...	...	...		5061	...	...	...
B	Mining and quarrying		267426	...	...	...		120210	...	...	...
35	Electricity, gas, steam & air conditioning		12635557	...	...	...		10360529	...	...	...
351	Electric power generation, transmission		12635557	...	...	...		10360529	...	...	...
352	Manufacture of gas		...	...	...	...		...	...	...	...
353	Steam and air conditioning supply		...	...	...	...		...	...	...	...
D	Electricity, gas, steam & air conditioning		12635557a/	...	...	...		10360529a/	...	...	...
36	Water collection, treatment and supply		354939	...	...	...		247571	...	...	...
37	Sewerage		12182	...	...	...		11291	...	...	...
38	Waste collection,treatment,disposal activities		129129	...	...	...		24321	...	...	...
381	Waste collection		41494	...	...	...		6019	...	...	...
382	Waste treatment and disposal		67443	...	...	...		10551	...	...	...
383	Materials recovery		20191	...	...	...		7750	...	...	...
39	Remediation activities		...	...	...	...		...	...	...	...
E	Water supply;sewerage,waste management		496249a/	...	...	...		283182a/	...	...	...

a/ Sum of available data.

Poland

ISIC Revision 4		Number of enterprises					Number of employees					Wages and salaries paid to employees				
			(number)					(number)					(millions of Polish Zlotys)			
ISIC	Industry	Note	2010	2011	2012	2013	Note	2010	2011	2012	2013	Note	2010	2011	2012	2013
051	Mining of hard coal		41	46	51	51		119383	116683	116783	112104		8470	9320	9602	9422
052	Mining of lignite		7	7	8	6		5497	5346	5092	4619		317	327	325	269
061	Extraction of crude petroleum		31	26	24	26		...	...	...	...		...	...	...	...
062	Extraction of natural gas		23	38	29	26		...	...	82	130		...	...	...	...
071	Mining of iron ores		1	2	-	-		...	...	-	-		...	...	-	-
072	Mining of non-ferrous metal ores		8	10	11	12		...	...	...	...		...	...	...	...
081	Quarrying of stone, sand and clay		1365	1519	1433	1177		18241	18136	17214	15605		820	867	801	721
089	Mining and quarrying n.e.c.		105	101	94	90		4134	4154	4409	4143		211	242	238	241
091	Support activities for petroleum		90	101	102	92		4009	3818	4429	5047		309	289	350	436
099	Support activities for other mining, quarrying		114	164	192	177		3655	4464	5933	8893		137	190	255	373
B	Mining and quarrying		1785	2014	1944	1657		174342	171913	173068	169764		12279	13332	13785	13697
351	Electric power generation, transmission		1393	1787	2015	1933		...	...	87213	81674		...	...	6868	6621
352	Manufacture of gas		68	88	92	74		...	...	...	...		...	...	...	...
353	Steam and air conditioning supply		586	628	623	539		38412	37895	...	...		2035	2085	...	...
D	Electricity, gas, steam & air conditioning		2047	2503	2730	2546		160276	154249	145324	137505		10281	10436	10542	10106
360	Water collection, treatment and supply		533	619	627	658		32688	31606	31620	32023		1388	1381	1457	1507
370	Sewerage		1859	1926	2081	2103		31632	32445	32351	33497		1297	1445	1486	1568
381	Waste collection		1717	1937	2063	1939		32864	35129	35307	36220		1186	1292	1347	1419
382	Waste treatment and disposal		291	336	358	345		7588	7425	8568	9626		291	306	352	401
383	Materials recovery		1216	1291	1414	1406		8656	9402	9286	8690		310	361	349	309
390	Remediation activities		148	196	246	233		1056	799	1275	1142		35	29	48	41
E	Water supply;sewerage,waste management		5764	6305	6789	6684		114484	116806	118407	121198		4507	4815	5039	5245

Poland

ISIC Revision 4		Output (valuation not defined)					Value added at factor values				
			(millions of Polish Zlotys)					(millions of Polish Zlotys)			
ISIC	Industry	Note	2010	2011	2012	2013	Note	2010	2011	2012	2013
051	Mining of hard coal		25163	29447	30081	26961		17152	20303	19842	17044
052	Mining of lignite		978	1043	1017	1060		684	690	583	664
061	Extraction of crude petroleum		...	...	...	...		...	...	...	...
062	Extraction of natural gas		...	...	71	130		...	...	15	-6
071	Mining of iron ores		...	...	-	-		...	...	-	-
072	Mining of non-ferrous metal ores		...	...	...	...		...	...	...	...
081	Quarrying of stone, sand and clay		6218	7731	6989	5999		1731	3116	2307	1836
089	Mining and quarrying n.e.c.		1043	1659	1430	1306		452	613	708	590
091	Support activities for petroleum		1421	1331	1319	1794		639	672	565	785
099	Support activities for other mining, quarrying		518	625	847	1458		208	335	429	981
B	Mining and quarrying		51975	62880	63549	57206		31425	39350	37694	32796
351	Electric power generation, transmission		...	...	71322	68711		...	...	37023	36376
352	Manufacture of gas		...	...	...	...		...	...	...	...
353	Steam and air conditioning supply		14466	14096	...	...		6664	6711	...	...
D	Electricity, gas, steam & air conditioning		107083	109103	116008	115441		47204	48696	52250	51572
360	Water collection, treatment and supply		5815	5915	6250	6452		3815	3857	4121	4307
370	Sewerage		5978	6556	6976	8654		3848	4473	4653	6157
381	Waste collection		6569	6838	7160	7790		3021	3160	3107	3204
382	Waste treatment and disposal		1900	1973	2118	2311		985	981	1022	1140
383	Materials recovery		4656	5765	5459	4681		1062	1197	991	766
390	Remediation activities		248	266	440	372		91	76	116	83
E	Water supply;sewerage,waste management		25166	27313	28403	30261		12823	13745	14009	15658

Portugal

ISIC Revision 4		Number of enterprises					Number of employees					Wages and salaries paid to employees				
		Note	(number)				Note	(number)				Note	(millions of Euros)			
ISIC	Industry		2010	2011	2012	2013		2010	2011	2012	2013		2010	2011	2012	2013
05	Mining of coal and lignite		-	-	-	-		-	-	-	-		-	-	-	-
051	Mining of hard coal		-	-	-	-		-	-	-	-		-	-	-	-
052	Mining of lignite		-	-	-	-		-	-	-	-		-	-	-	-
06	Extraction of crude petroleum & natural gas		-	-	-	-		-	-	-	-		-	-	-	-
061	Extraction of crude petroleum		-	-	-	-		-	-	-	-		-	-	-	-
062	Extraction of natural gas		-	-	-	-		-	-	-	-		-	-	-	-
07	Mining of metal ores		19	21	18	19		...	...	...	...		...	...	...	...
071	Mining of iron ores		4	5	4	3		...	...	...	...		...	...	...	...
072	Mining of non-ferrous metal ores		15	16	14	16		1431	...	...	...		36	...	...	...
08	Other mining and quarrying		1276	1227	1138	1119		9642	8869	7664	7027		125	115	98	88
081	Quarrying of stone, sand and clay		1201	1149	1054	1026		9331	8576	7374	6759		121	111	94	85
089	Mining and quarrying n.e.c.		75	78	84	93		311	293	290	268		4	4	4	3
09	Mining support service activities		26	26	21	19		...	...	...	...		...	...	...	...
091	Support activities for petroleum		1	1	-	2		...	...	-	...		...	...	-	...
099	Support activities for other mining, quarrying		25	25	21	17		...	...	...	366		...	...	...	7
B	Mining and quarrying		1321	1274	1177	1157		11265	10636	9643	9091		163	163	153	145
35	Electricity, gas, steam & air conditioning		730	801	881	925		8822	8609	8483	8163		336	318	291	317
351	Electric power generation, transmission		663	734	820	867		8013	7774	7705	7437		306	290	262	291
352	Manufacture of gas		44	45	41	41		731	758	704	652		28	27	27	25
353	Steam and air conditioning supply		23	22	20	17		78	77	74	74		2	2	2	2
D	Electricity, gas, steam & air conditioning		730	801	881	925		8822	8609	8483	8163		336	318	291	317
36	Water collection, treatment and supply		144	152	151	142		12999	13568	13475	13194		221	210	198	213
37	Sewerage		71	69	72	71		1605	1646	1679	1675		30	28	26	29
38	Waste collection,treatment,disposal activities		834	907	932	985		14842	15086	14770	14567		211	204	194	198
381	Waste collection		277	326	339	359		6179	6113	5932	5995		79	76	72	73
382	Waste treatment and disposal		103	108	110	113		5322	5457	5325	5188		85	82	76	81
383	Materials recovery		454	473	483	513		3341	3516	3513	3384		46	46	46	44
39	Remediation activities		20	21	18	26		65	60	67	71		1	1	1	1
E	Water supply;sewerage,waste management		1069	1149	1173	1224		29511	30360	29991	29507		463	443	418	441

Portugal

ISIC Revision 4		Output (valuation not defined)					Value added at factor values				
			(millions of Euros)					(millions of Euros)			
ISIC	Industry	Note	2010	2011	2012	2013	Note	2010	2011	2012	2013
05	Mining of coal and lignite		-	-	-	-		-	-	-	-
051	Mining of hard coal		-	-	-	-		-	-	-	-
052	Mining of lignite		-	-	-	-		-	-	-	-
06	Extraction of crude petroleum & natural gas		-	-	-	-		-	-	-	-
061	Extraction of crude petroleum		-	-	-	-		-	-	-	-
062	Extraction of natural gas		-	-	-	-		-	-	-	-
07	Mining of metal ores		...	...	...	...		...	...	...	...
071	Mining of iron ores		...	...	...	...		...	...	...	...
072	Mining of non-ferrous metal ores		453	...	...	...		280	...	...	...
08	Other mining and quarrying		722	682	537	501		270	238	185	182
081	Quarrying of stone, sand and clay		694	655	518	482		259	229	179	175
089	Mining and quarrying n.e.c.		28	27	19	19		11	9	6	7
09	Mining support service activities		...	...	...	...		...	...	...	...
091	Support activities for petroleum		...	...	-	...		...	...	-	...
099	Support activities for other mining, quarrying		...	...	...	49		...	...	...	28
B	Mining and quarrying		1189	1233	1090	1028		557	520	452	419
35	Electricity, gas, steam & air conditioning		12575	13258	13125	13005		3896	3903	4122	4382
351	Electric power generation, transmission		12017	12663	12540	12118		3638	3623	3846	3920
352	Manufacture of gas		542	578	567	868		250	273	269	454
353	Steam and air conditioning supply		16	17	18	19		7	7	7	8
D	Electricity, gas, steam & air conditioning		12575	13258	13125	13005		3896	3903	4122	4382
36	Water collection, treatment and supply		1203	1318	1325	1316		641	709	705	707
37	Sewerage		209	231	237	259		115	126	134	147
38	Waste collection,treatment,disposal activities		1342	1318	1263	1193		552	532	506	483
381	Waste collection		328	321	319	308		159	151	148	143
382	Waste treatment and disposal		527	506	480	482		249	244	235	238
383	Materials recovery		487	492	464	403		145	137	123	102
39	Remediation activities		3	2	2	2		1	1	1	1
E	Water supply;sewerage,waste management		2756	2869	2827	2770		1310	1367	1346	1337

Qatar

ISIC Revision 3		Number of establishments					Number of employees					Wages and salaries paid to employees				
			(number)					(number)					(millions of Qatari Riyals)			
ISIC	Industry	Note	2010	2011	2012	2013	Note	2010	2011	2012	2013	Note	2010	2011	2012	2013
101	Mining and agglomeration of hard coal		...	...	...	...		...	...	...	...		...	...	...	...
102	Mining and agglomeration of lignite		...	...	...	...		...	...	...	...		...	...	...	...
103	Extraction and agglomeration of peat		...	...	...	...		...	...	...	...		...	...	...	...
111	Extraction of crude oil and natural gas		7	7	7	7		19103	19351	19857	19556		7296	7995	9289	9921
112	Oil, gas service activities		70	76	90	78		35251	29672	18722	20731		2547	1607	1921	2386
120	Mining of uranium and thorium ores		...	...	...	...		...	...	...	...		...	...	...	...
131	Iron ores		...	...	...	...		...	...	...	...		...	...	...	...
132	Non-ferrous metal ores,excl.uranium,thorium		...	...	...	...		...	...	...	...		...	...	...	...
141	Quarrying of stone, sand and clay		10	8	5	5		866	598	511	522		40	20	19	23
142	Mining and quarrying n.e.c.		...	...	...	...		...	...	...	...		...	...	...	...
C	Total mining and quarrying		87	91	102	90		55220	49621	39090	40809		9882	9622	11230	12329
401	Electricity production, collection, distribution		...	...	...	...		...	...	...	...		...	...	...	...
402	Gas production and distribution		...	...	...	...		...	...	...	...		...	...	...	...
403	Steam and hot water supply		...	...	...	...		...	...	...	...		...	...	...	...
410	Water collection, purification, distribution		...	...	...	...		...	...	...	...		...	...	...	...
E	Total electricity, gas and water supply		7	7	7	7		4919	4889	4800	4726		907	1025	1016	1260

Qatar

ISIC Revision 3		Output at producers' prices					Value added at producers' prices				
		Note	(millions of Qatari Riyals)				Note	(millions of Qatari Riyals)			
ISIC	Industry		2010	2011	2012	2013		2010	2011	2012	2013
101	Mining and agglomeration of hard coal		...	...	...	...		...	...	...	...
102	Mining and agglomeration of lignite		...	...	...	...		...	...	...	...
103	Extraction and agglomeration of peat		...	...	...	...		...	...	...	...
111	Extraction of crude oil and natural gas		260779	402813	425336	431372		229786	345198	388661	392683
112	Oil, gas service activities		9115	9126	7948	9735		6873	6253	4955	6688
120	Mining of uranium and thorium ores		...	...	...	...		...	...	...	...
131	Iron ores		...	...	...	...		...	...	...	...
132	Non-ferrous metal ores,excl.uranium,thorium		...	...	...	...		...	...	...	...
141	Quarrying of stone, sand and clay		272	252	241	275		151	207	118	165
142	Mining and quarrying n.e.c.		...	...	...	...		...	...	...	...
C	Total mining and quarrying		270166	412191	433525	441383		236811	351658	393735	399537
401	Electricity production, collection, distribution		...	...	...	...		...	...	...	...
402	Gas production and distribution		...	...	...	...		...	...	...	...
403	Steam and hot water supply		...	...	...	...		...	...	...	...
410	Water collection, purification, distribution		...	...	...	...		...	...	...	...
E	Total electricity, gas and water supply		8791	12543	13672	14795		1438	2932	3360	3946

Republic of Korea

ISIC Revision 4		Number of establishments					Number of employees					Wages and salaries paid to employees				
			(number)					(number)					(billions of Korean Won)			
ISIC	Industry	Note	2009	2010	2011	2012	Note	2009	2010	2011	2012	Note	2009	2010	2011	2012
051	Mining of hard coal	a/	5	7	6	6	a/	3630	3206	3115	2867	a/	149	147	144	150
052	Mining of lignite	a/	...	...	...	...	a/	...	...	...	...	a/	...	...	...	...
061	Extraction of crude petroleum	b/	1	1	1	1		...	67b/	...	...		...	5b/	...	...
062	Extraction of natural gas	b/	...	...	...	...		...	...b/	...	...		...	...b/	...	...
071	Mining of iron ores		3c/	2	6c/	7c/		174c/	133	293c/	353c/		5c/	4	9c/	14c/
072	Mining of non-ferrous metal ores		...c/	14	...c/	...c/		...c/	151	...c/	...c/		...c/	3	...c/	...c/
081	Quarrying of stone, sand and clay		371d/	663	344d/	312d/		8366d/	9171	7410d/	6840d/		250d/	276	245d/	236d/
089	Mining and quarrying n.e.c.		...d/	1075	...d/	...d/		...d/	1247	...d/	...d/		...d/	20	...d/	...d/
091	Support activities for petroleum	e/	1	8	2	2		...	414e/	...	...		...	21e/	...	...
099	Support activities for other mining, quarrying	e/	...	...	...	...		...	...e/	...	...		...	...e/	...	...
B	Mining and quarrying		381	1770	359	328		...	14389	...	...		...	476	...	...
351	Electric power generation, transmission		...	934f/	...	...		...	50897f/	...	...		...	3128f/	...	...
352	Manufacture of gas		...	...f/	...	...		...	...f/	...	...		...	...f/	...	...
353	Steam and air conditioning supply		...	...f/	...	...		...	...f/	...	...		...	...f/	...	...
D	Electricity, gas, steam & air conditioning		...	934	...	...		...	50897	...	...		...	3128	...	...
360	Water collection, treatment and supply		...	...	...	...		...	...	...	...		...	...	...	...
370	Sewerage		...	...	1356	1485		13052	...	13882	14763		414	...	451	503
381	Waste collection		1328	1441	1637	1881		20013	20992	21618	22167		537	580	577	618
382	Waste treatment and disposal		948	2601	1088	1163		17992	30905	18998	19169		482	834	567	606
383	Materials recovery		1485	1584	1744	1910		11237	12229	12852	13204		244	295	285	370
390	Remediation activities		...	...	84	85		965	...	972	973		21	...	26	30
E	Water supply;sewerage,waste management		...	7551	...	...		...	94180	...	...		...	2812	...	...

a/ 051 includes 052.
b/ 061 includes 062.
c/ 071 includes 072.
d/ 081 includes 089.
e/ 091 includes 099.
f/ 351 includes 352 and 353.

Republic of Korea

ISIC Revision 4		Output (valuation not defined)					Value added (valuation not defined)				
			(billions of Korean Won)					(billions of Korean Won)			
ISIC	Industry	Note	2009	2010	2011	2012	Note	2009	2010	2011	2012
051	Mining of hard coal	a/	308	294	294	317		210a/	...	201a/	223a/
052	Mining of lignite	a/	...	...	...	...		...a/	...	...a/	...a/
061	Extraction of crude petroleum		...	225b/	...	...		...	...	...	...
062	Extraction of natural gas		...	...b/	...	...		...	...	...	...
071	Mining of iron ores		41c/	40	92c/	82c/		34c/	...	77c/	73c/
072	Mining of non-ferrous metal ores		...c/	32	...c/	...c/		...c/	...	...c/	...c/
081	Quarrying of stone, sand and clay		2260d/	2784	2256d/	2126d/		1476d/	...	1472d/	1437d/
089	Mining and quarrying n.e.c.		...d/	115	...d/	...d/		...d/	...	...d/	...d/
091	Support activities for petroleum		...	165e/	...	...		...	...	...	...
099	Support activities for other mining, quarrying		...	...e/	...	...		...	...	...	...
B	Mining and quarrying		...	3655	...	...		...	...	...	...
351	Electric power generation, transmission		...	113578f/	...	...		...	...	...	...
352	Manufacture of gas		...	...f/	...	...		...	...	...	...
353	Steam and air conditioning supply		...	...f/	...	...		...	...	...	...
D	Electricity, gas, steam & air conditioning		...	113578	...	...		...	...	...	...
360	Water collection, treatment and supply		...	...	...	...		...	...	...	...
370	Sewerage		...	...	2407	2965		674	...	824	1046
381	Waste collection		1847	2358	2494	2277		872	...	986	1054
382	Waste treatment and disposal		3029	9712	3535	3380		1291	...	1366	1370
383	Materials recovery		3760	6485	7725	7601		657	...	828	906
390	Remediation activities		...	...	170	164		46	...	50	56
E	Water supply;sewerage,waste management		...	27938	...	...		...	...	...	...

a/ 051 includes 052.
b/ 061 includes 062.
c/ 071 includes 072.
d/ 081 includes 089.
e/ 091 includes 099.
f/ 351 includes 352 and 353.

Romania

ISIC Revision 4		Number of establishments					Number of employees					Wages and salaries paid to employees				
			(number)					(number)					(millions of Romanian Lei)			
ISIC	Industry	Note	2010	2011	2012	2013	Note	2010	2011	2012	2013	Note	2010	2011	2012	2013
05	Mining of coal and lignite		62	60	75	57		18005	17067	11482	5754		...	...	134	233
051	Mining of hard coal		24	20	27	18		9312	8508	7719	5373		...	...	...	...
052	Mining of lignite		38	40	48	39		8693	8559	3763	381		...	...	...	...
06	Extraction of crude petroleum & natural gas		33	55	76	77		30541	28048	26547	25677		...	...	...	...
061	Extraction of crude petroleum		28	43	52	56		24605	22086	20528	19097		...	...	...	...
062	Extraction of natural gas		5	12	24	21		5936	5962	6019	6580		...	...	...	...
07	Mining of metal ores		35	34	45	42		2500	2562	2672	2584		...	...	...	...
071	Mining of iron ores		2	3	7	2		69	76	71	36		...	...	...	...
072	Mining of non-ferrous metal ores		33	31	38	40		2431	2486	2601	2548		...	...	...	...
08	Other mining and quarrying		1079	1023	1002	972		10252	10111	10420	9830		205	209	214	229
081	Quarrying of stone, sand and clay		1040	985	944	914		7741	7626	7934	7431		...	...	...	...
089	Mining and quarrying n.e.c.		39	38	58	58		2511	2485	2486	2399		...	...	...	...
09	Mining support service activities		141	142	138	137		6464	7218	6904	7800		250	283	301	428
091	Support activities for petroleum		112	115	114	113		6255	6924	6622	7508		246	277	295	418
099	Support activities for other mining, quarrying		29	27	24	24		209	294	282	292		4	6	6	10
B	Mining and quarrying	a/	1350	1314	1336	1285	a/	67762	65006	58025	51645		3041	2960	2814	2790
35	Electricity, gas, steam & air conditioning		1047	1068	1192	1467		80844	79239	79304	76916		3585	3580	3772	3645
351	Electric power generation, transmission		843	848	984	1254		50915	51241	53660	55517		2551	2528	2765	2753
352	Manufacture of gas		102	112	95	100		10559	10307	10002	7158		491	508	514	400
353	Steam and air conditioning supply		102	108	113	113		19370	17691	15642	14241		543	544	493	492
D	Electricity, gas, steam & air conditioning	a/	1047	1068	1192	1467	a/	80844	79239	79304	76916	a/	3585	3580	3772	3645
36	Water collection, treatment and supply		377	408	431	465		34884	35347	36002	36648		806	920	969	1036
37	Sewerage		184	183	194	211		1034	1116	1460	1437		26	35	40	39
38	Waste collection,treatment,disposal activities		2165	2321	2585	2674		42403	47651	49072	48522		689	798	873	892
381	Waste collection		688	811	924	985		28714	30815	32266	32101		464	527	592	599
382	Waste treatment and disposal		92	103	109	129		1282	1728	1813	2278		29	39	41	53
383	Materials recovery		1385	1407	1552	1560		12407	15108	14993	14143		196	232	240	240
39	Remediation activities		58	58	63	63		650	551	601	569		14	15	15	13
E	Water supply;sewerage,waste management	a/	2784	2970	3273	3413	a/	78971	84665	87135	87176	a/	1535	1768	1897	1980

a/ Sum of available data.

Romania

ISIC Revision 4		Output at factor values a/					Value added at factor values				
			(millions of Romanian Lei)					(millions of Romanian Lei)			
ISIC	Industry	Note	2010	2011	2012	2013	Note	2010	2011	2012	2013
05	Mining of coal and lignite		1508	1745	1050	451		...	...	307	418
051	Mining of hard coal		463	487	495	359		...	...	...	...
052	Mining of lignite		1045	1258	555	92		...	...	...	...
06	Extraction of crude petroleum & natural gas		17655	20972	23641	22323		...	...	...	...
061	Extraction of crude petroleum		13997	16678	19711	18233		...	...	...	...
062	Extraction of natural gas		3658	4294	3930	4090		...	...	...	...
07	Mining of metal ores		291	329	337	365		...	...	...	...
071	Mining of iron ores		10	12	3	2		...	...	...	...
072	Mining of non-ferrous metal ores		281	317	334	363		...	...	...	...
08	Other mining and quarrying		1760	1816	2040	1851		571	529	582	853
081	Quarrying of stone, sand and clay		1457	1506	1689	1400		...	...	...	...
089	Mining and quarrying n.e.c.		303	310	351	451		...	...	...	...
09	Mining support service activities		3732	2593	2799	3573		760	566	768	927
091	Support activities for petroleum		3681	2546	2737	3519		740	547	755	887
099	Support activities for other mining, quarrying		51	47	62	54		20	19	13	40
B	Mining and quarrying	b/	24946	27455	29867	28563		12376	14745	16069	15927
35	Electricity, gas, steam & air conditioning		50874	57012	59718	55996		13832	12978	15098	20341
351	Electric power generation, transmission		38879	42011	42781	39905		10985	10560	12565	17417
352	Manufacture of gas		9543	12329	14315	13421		1494	1352	1864	2262
353	Steam and air conditioning supply		2452	2672	2622	2670		1353	1066	669	663
D	Electricity, gas, steam & air conditioning	b/	50874	57012	59718	55996	b/	13832	12978	15098	20341
36	Water collection, treatment and supply		2372	2738	3167	3403		1789	1715	2068	2057
37	Sewerage		213	199	229	208		66	84	91	79
38	Waste collection,treatment,disposal activities		11938	13108	13042	11448		1979	2135	2055	2131
381	Waste collection		2204	2431	2679	2839		998	1138	1171	1273
382	Waste treatment and disposal		404	458	326	461		145	146	121	217
383	Materials recovery		9330	10219	10037	8148		836	851	763	641
39	Remediation activities		118	211	235	124		42	91	98	47
E	Water supply;sewerage,waste management	b/	14641	16256	16673	15183	b/	3876	4025	4312	4314

a/ Output refers to turnover.
b/ Sum of available data.

Russian Federation

ISIC Revision 3		Number of establishments					Number of employees					Wages and salaries paid to employees a/				
			(number)					(thousands)					(millions of Russian Roubles)			
ISIC	Industry	Note	2010	2011	2012	2013	Note	2010	2011	2012	2013	Note	2010	2011	2012	2013
10	Mining of coal and lignite; extraction of peat		448	756	674	828		167	168	168	163		54079	65060	72054	74457
101	Mining and agglomeration of hard coal		277	464	494	516		142	145	146	142		47776	57983	64502	66734
102	Mining and agglomeration of lignite		57	87	90	108		21	20	20	19		5748	6581	7010	7165
103	Extraction and agglomeration of peat		114	205	90	204		4	3	3	3		554	497	542	558
11	Extraction of crude oil and natural gas, etc.		1391	2086	2255	2243		405	412	429	434		265199	296103	338967	364286
111	Extraction of crude oil and natural gas		522	775	820	737		187	184	185	179		135783	149603	166028	175025
112	Oil, gas service activities		869	1311	1435	1506		218	228	244	255		129416	146500	172939	189261
12	Mining of uranium and thorium ores		...	...	...	...		...	...	...	...		...	...	...	...
120	Mining of uranium and thorium ores	b/	...	...	...	...	b/	...	...	...	...	b/	...	...	...	...
13	Mining of metal ores		606	1133	349	1246		182	190	196	195		71316	87087	100344	106814
131	Iron ores		50	88	32	95		62	63	63	62		19218	23031	25632	27409
132	Non-ferrous metal ores,excl.uranium,thorium		556	1045	317	1151		121	127	133	132		52099	64056	74712	79405
14	Other mining and quarrying		2334	3819	4263	4464		90	93	94	96		18717	22761	25875	28921
141	Quarrying of stone, sand and clay		2182	3510	3928	4153		42	45	44	43		8785	11384	12805	13668
142	Mining and quarrying n.e.c.		152	309	335	311		48	48	50	54		9932	11377	13071	15252
C	Total mining and quarrying		4784	7803	7547	8786		898	917	886	888		429861	496803	537241	574478
40	Electricity, gas, steam and hot water supply		10682	13569	14511	14502		1546	1558	1556	1558		474357	534470	581882	636320
401	Electricity production, collection, distribution		3447	4809	5327	5463		702	713	713	717		276380	314485	340772	373447
402	Gas production and distribution		488	631	703	581		163	161	161	166		45239	50221	54496	60723
403	Steam and hot water supply		6747	8129	8481	8458		682	684	682	676		152738	169764	186615	202151
41	Water collection, purification, distribution		3219	3981	4312	4371		292	286	277	271		58598	62422	65642	70994
410	Water collection, purification, distribution		3219	3981	4312	4371		292	286	277	271		58598	62422	65642	70994
E	Total electricity, gas and water supply		13901	17550	18823	18873		1839	1845	1833	1829		532959	596908	647532	707329

a/ Data were computed from the reported monthly average wages.

b/ Data suppressed due to confidentiality rules.

Russian Federation

ISIC Revision 3		Output (valuation not defined) a/					Value added (valuation not defined) b/				
			(billions of Russian Roubles)					(billions of Russian Roubles)			
ISIC	Industry	Note	2010	2011	2012	2013	Note	2010	2011	2012	2013
10	Mining of coal and lignite; extraction of peat		508.4	680.6	679.6	636.2		264.6	347.5	280.6	209.1
101	Mining and agglomeration of hard coal		474.8	642.0	631.6	583.9		246.2	326.7	255.9	182.9
102	Mining and agglomeration of lignite		31.7	36.6	45.5	50.1		17.6	19.9	23.6	25.3
103	Extraction and agglomeration of peat		1.8	2.0	2.5	2.2		0.8	0.9	1.1	0.9
11	Extraction of crude oil and natural gas, etc.		4616.4	5922.6	7013.7	7356.6		3096.3	4118.2	4940.0	5089.5
111	Extraction of crude oil and natural gas		3989.9	5191.2	6078.1	6309.7		2882.7	3882.3	4622.0	4726.7
112	Oil, gas service activities		626.5	731.4	935.5	1046.9		213.5	236.0	318.1	362.8
12	Mining of uranium and thorium ores		...	...	...	...		...	...	...	...
120	Mining of uranium and thorium ores	c/	...	...	...	...	c/	...	...	...	...
13	Mining of metal ores		577.3	809.9	721.6	710.4		336.3	494.7	381.9	358.7
131	Iron ores		255.2	365.1	312.0	323.9		154.3	238.9	175.4	178.4
132	Non-ferrous metal ores,excl.uranium,thorium		322.1	444.8	409.6	386.5		182.1	255.8	206.5	180.3
14	Other mining and quarrying		268.1	349.1	412.0	428.4		139.8	191.2	220.8	234.5
141	Quarrying of stone, sand and clay	d/	268.1	349.1	412.0	428.4	d/	139.8	191.2	220.8	234.5
142	Mining and quarrying n.e.c.	d/	...	...	...	...	d/	...	...	...	...
C	Total mining and quarrying		5982.2	7774.2	8831.1	9135.1		3842.8	5157.3	5826.1	5893.9
40	Electricity, gas, steam and hot water supply		4293.2	4935.3	5045.2	5285.8		1405.7	1678.6	1686.4	1830.0
401	Electricity production, collection, distribution		3269.0	3731.5	3762.6	3923.2		1083.4	1280.5	1296.9	1429.4
402	Gas production and distribution		199.4	235.0	288.9	300.6		107.6	129.9	142.0	138.0
403	Steam and hot water supply		824.8	968.8	993.6	1062.0		214.7	268.3	247.5	262.6
41	Water collection, purification, distribution		243.7	266.6	270.2	280.5		121.4	135.8	137.8	145.6
410	Water collection, purification, distribution		243.7	266.6	270.2	280.5		121.4	135.8	137.8	145.6
E	Total electricity, gas and water supply		4536.9	5201.9	5315.4	5566.2		1527.1	1814.5	1824.2	1975.6

a/ Output is reported at basic prices.
b/ Value added is reported at basic prices.
c/ Data suppressed due to confidentiality rules.
d/ 141 includes 142.

Senegal

ISIC Revision 4		Number of establishments					Employees					Wages and salaries				
			(number)					(number)					(millions of West African CFA Francs)			
ISIC	Industry	Note	2009	2010	2011	2012	Note	2009	2010	2011	2012	Note	2009	2010	2011	2012
05	Mining of coal and lignite		...	...	...	...		...	...	...	...		...	...	...	...
051	Mining of hard coal		...	...	...	...		...	...	...	...		...	...	...	...
052	Mining of lignite		...	...	...	...		...	...	...	...		...	...	...	...
06	Extraction of crude petroleum & natural gas		2	2	2	2		...	...	...	...		...	...	...	...
061	Extraction of crude petroleum		2	2	2	2		...	...	...	...		...	...	...	...
062	Extraction of natural gas		...	...	...	...		...	...	...	...		...	...	...	...
07	Mining of metal ores		3	4	5	5		...	...	...	...		...	...	...	...
071	Mining of iron ores		1	1	2	2		...	...	...	...		...	...	...	...
072	Mining of non-ferrous metal ores		2	3	3	3		...	...	...	...		...	...	...	...
08	Other mining and quarrying		24	25	25	25		...	...	...	...		...	...	...	...
081	Quarrying of stone, sand and clay		12	12	10	10		...	...	...	...		...	...	...	...
089	Mining and quarrying n.e.c.		12	13	15	15		...	...	...	...		...	...	...	...
09	Mining support service activities		9	9	9	10		...	...	...	...		...	...	...	...
091	Support activities for petroleum		8	8	8	8		...	...	...	...		...	...	...	...
099	Support activities for other mining, quarrying		1	1	1	2		...	...	...	...		...	...	...	...
B	Mining and quarrying	a/	38	40	41	42		...	...	...	...		...	...	...	...
35	Electricity, gas, steam & air conditioning		32	33	37	37		...	...	...	...		...	...	...	...
351	Electric power generation, transmission		8	8	8	8		...	...	...	...		...	...	...	...
352	Manufacture of gas		3	3	4	4		...	...	...	...		...	...	...	...
353	Steam and air conditioning supply		21	22	25	25		...	...	...	...		...	...	...	...
D	Electricity, gas, steam & air conditioning	a/	32	33	37	37		...	...	...	...		...	...	...	...
36	Water collection, treatment and supply		14	13	13	13		...	...	...	...		...	...	...	...
37	Sewerage		7	6	8	8		...	...	...	...		...	...	...	...
38	Waste collection,treatment,disposal activities		6	6	6	7		...	...	...	...		...	...	...	...
381	Waste collection		6	6	6	7		...	...	...	...		...	...	...	...
382	Waste treatment and disposal		...	...	...	...		...	...	...	...		...	...	...	...
383	Materials recovery		...	...	...	...		...	...	...	...		...	...	...	...
39	Remediation activities		...	...	...	...		...	...	...	...		...	...	...	...
E	Water supply;sewerage,waste management	a/	27	25	27	28		...	...	...	...		...	...	...	...

a/ Sum of available data.

Senegal

ISIC Revision 4		Output at producers' prices					Value added at producers' prices				
			(millions of West African CFA Francs)					(millions of West African CFA Francs)			
ISIC	Industry	Note	2009	2010	2011	2012	Note	2009	2010	2011	2012
05	Mining of coal and lignite		...	...	...	...		...	...	...	...
051	Mining of hard coal		...	...	...	...		...	...	...	...
052	Mining of lignite		...	...	...	...		...	...	...	...
06	Extraction of crude petroleum & natural gas		1957	5134	3208	5005		641	2935	108	-607
061	Extraction of crude petroleum		1957	5134	3208	5005		641	2935	108	-607
062	Extraction of natural gas		...	...	...	...		...	...	...	...
07	Mining of metal ores		1061	7348	7618	10075		173	1026	-504	-752
071	Mining of iron ores		544	1216	3075	3352		43	110	-389	-1756
072	Mining of non-ferrous metal ores		517	6132	4543	6723		130	916	-115	1004
08	Other mining and quarrying		45101	51573	63521	65334		-5269	-4707	-1077	-5238
081	Quarrying of stone, sand and clay		25838	29814	38991	40954		6817	7680	11738	11507
089	Mining and quarrying n.e.c.		19263	21759	24530	24380		-12086	-12387	-12815	-16745
09	Mining support service activities		97184	122131	129946	135683		19621	49748	53082	54748
091	Support activities for petroleum		97184	122131	129946	135425		20032	50180	53534	55139
099	Support activities for other mining, quarrying		...	...	...	258		-411	-432	-452	-391
B	Mining and quarrying	a/	145303	186186	204293	216097	a/	15166	49002	51609	48151
35	Electricity, gas, steam & air conditioning		315231	333539	416602	433866		68407	12604	62473	65884
351	Electric power generation, transmission		294008	308177	388463	407208		64865	9231	59418	63217
352	Manufacture of gas		18726	22361	24908	23327		3041	2877	2615	2253
353	Steam and air conditioning supply		2497	3001	3231	3331		501	496	440	414
D	Electricity, gas, steam & air conditioning	a/	315231	333539	416602	433866	a/	68407	12604	62473	65884
36	Water collection, treatment and supply		63757	78268	71923	80660		17264	13745	16959	17974
37	Sewerage		1349	1467	2000	2872		74	101	220	264
38	Waste collection,treatment,disposal activities		11599	12133	11504	12450		3502	3988	2335	3489
381	Waste collection		11599	12133	11504	12450		3502	3988	2335	3489
382	Waste treatment and disposal		...	...	...	...		...	...	...	...
383	Materials recovery		...	...	...	...		...	...	...	...
39	Remediation activities		...	...	...	...		...	...	...	...
E	Water supply;sewerage,waste management	a/	76705	91868	85427	95982	a/	20840	17834	19514	21727

a/ Sum of available data.

Serbia

ISIC Revision 4		Output (valuation not defined) a/					Value added (valuation not defined) b/				
			(millions of Serbian Dinars)					(millions of Serbian Dinars)			
ISIC	Industry	Note	2009	2010	2011	2012	Note	2009	2010	2011	2012
05	Mining of coal and lignite		43243	49098	56694	58572		29005	31959	38164	38132
06	Extraction of crude petroleum & natural gas		9105	8848	9925	11361		2364	2737	3869	4811
07	Mining of metal ores		7145	10860	15575	23341		1376	4734	7478	11071
08	Other mining and quarrying		10832	9874	10028	11789		3643	2906	3108	2982
09	Mining support service activities		-	-	-	-		-	-	-	-
B	Mining and quarrying		70324	78680	92221	105063		36388	42336	52619	56996
35	Electricity, gas, steam & air conditioning		370054	407781	466842	487451		86388	90120	113845	109772
D	Electricity, gas, steam & air conditioning		370054	407781	466842	487451		86388	90120	113845	109772
36	Water collection, treatment and supply		25380	27810	31635	35257		15572	16707	18849	20494
37	Sewerage		-	12	211	...		-	6	88	...
38	Waste collection,treatment,disposal activities		32043	47595	59458	61777		13321	17030	19819	20186
39	Remediation activities		26	15	321	...		6	2	-31	...
E	Water supply;sewerage,waste management		57449	75432	91626	97034		28898	33744	38725	40680

a/ Output refers to gross output at basic prices.

b/ Value added refers to total value added at basic prices.

Slovakia

ISIC Revision 4		Number of enterprises					Number of employees					Wages and salaries paid to employees				
			(number)					(number)					(millions of Euros)			
ISIC	Industry	Note	2010	2011	2012	2013	Note	2010	2011	2012	2013	Note	2010	2011	2012	2013
051	Mining of hard coal		-	-	...	...		-	-	...	...		-	-	...	...
052	Mining of lignite		...	...	...	...		...	...	...	...		...	...	...	...
061	Extraction of crude petroleum		-	-	-	-		-	-	-	-		-	-	-	-
062	Extraction of natural gas		-	...	...	...		-	...	...	...		-	...	...	...
071	Mining of iron ores		-	-	-	-		-	-	-	-		-	-	-	-
072	Mining of non-ferrous metal ores		...	...	...	...		...	...	...	...		...	...	...	...
081	Quarrying of stone, sand and clay		63	95	81	106		1690	1843	1682	1820		18	21	21	21
089	Mining and quarrying n.e.c.		31	27	29	31		420	305	276	311		4	3	3	3
091	Support activities for petroleum		9a/	...	...	...		931a/	...	...	...		17a/	...	...	...
099	Support activities for other mining, quarrying		...a/	8	7	11		...a/	265	254	258		...a/	6	7	7
B	Mining and quarrying		108	138	125	157		7230	7381	7267	7332		86	90	91	91
351	Electric power generation, transmission		104	119	173	223		10192	9450	9550	9461		203	203	215	217
352	Manufacture of gas		8	8	8	12		3187	2837	2581	2442		65	69	62	59
353	Steam and air conditioning supply		182	130	177	195		6606	6406	6290	5790		82	81	81	76
D	Electricity, gas, steam & air conditioning		294	257	358	430		19985	18693	18421	17693		350	352	358	353
360	Water collection, treatment and supply		46	54	47	56		11690	11610	11623	11503		113	125	126	128
370	Sewerage		98	95	96	112		336	322	297	386		3	3	3	4
381	Waste collection		237	195	201	240		4315	4655	4280	4814		40	43	41	45
382	Waste treatment and disposal		117	118	123	155		2779	2809	2654	2797		29	39	36	36
383	Materials recovery		604	416	445	433		1427	1520	1760	1577		14	15	17	23
390	Remediation activities		26	28	17	19		285	246	225	196		2	3	2	2
E	Water supply;sewerage,waste management		1128	906	929	1015		20832	21162	20839	21273		201	228	226	238

a/ 091 includes 099.

Slovakia

ISIC Revision 4		Output (valuation not defined)					Value added at factor values				
		Note	(millions of Euros)				Note	(millions of Euros)			
ISIC	Industry		2010	2011	2012	2013		2010	2011	2012	2013
051	Mining of hard coal		-	-	...	...		-	-	...	...
052	Mining of lignite		...	...	...	...		...	...	...	...
061	Extraction of crude petroleum		-	-	-	-		-	-	-	-
062	Extraction of natural gas		-	...	...	...		-	...	...	...
071	Mining of iron ores		-	-	-	-		-	-	-	-
072	Mining of non-ferrous metal ores		...	...	...	...		...	...	...	...
081	Quarrying of stone, sand and clay		117	131	108	136		36	48	34	35
089	Mining and quarrying n.e.c.		17	14	11	14		6	4	4	5
091	Support activities for petroleum		223a/	...	...	...		171a/	...	...	...
099	Support activities for other mining, quarrying		...a/	15	14	15		...a/	10	10	10
B	Mining and quarrying		498	524	498	528		299	325	314	318
351	Electric power generation, transmission		6492	7496	8219	7698		1558	1644	1729	1548
352	Manufacture of gas		3404	3284	2901	2813		636	680	536	153
353	Steam and air conditioning supply		942	955	1060	1049		195	243	234	245
D	Electricity, gas, steam & air conditioning		10838	11735	12180	11560		2389	2567	2499	1945
360	Water collection, treatment and supply		469	460	482	486		252	244	269	266
370	Sewerage		20	19	20	24		8	5	9	10
381	Waste collection		200	177	186	264		108	90	96	136
382	Waste treatment and disposal		165	158	116	127		92	95	26	31
383	Materials recovery		94	115	131	172		42	54	51	44
390	Remediation activities		7	10	8	9		3	4	3	4
E	Water supply;sewerage,waste management		954	939	942	1080		505	491	453	491

a/ 091 includes 099.

Slovenia

ISIC Revision 4		Number of enterprises					Number of employees					Wages and salaries paid to employees				
			(number)					(number)					(millions of Euros)			
ISIC	Industry	Note	2010	2011	2012	2013	Note	2010	2011	2012	2013	Note	2010	2011	2012	2013
051	Mining of hard coal		-	-	-	-		-	-	-	-		-	-	-	-
052	Mining of lignite		2	1	1	1		...	...	...	...		...	...	...	...
061	Extraction of crude petroleum		-	-	-	-		-	-	-	-		-	-	-	-
062	Extraction of natural gas		1	1	1	1		...	-	-	-		...	...	...	...
071	Mining of iron ores		-	-	-	-		-	-	-	-		-	-	-	-
072	Mining of non-ferrous metal ores		-	-	-	-		-	-	-	-		-	-	-	-
081	Quarrying of stone, sand and clay		92	90	88	86		920	897	836	777		19	18	17	17
089	Mining and quarrying n.e.c.		7	12	10	12		...	...	174	168		...	...	...	...
091	Support activities for petroleum		3	3	3	4		...	...	...	68		...	...	...	...
099	Support activities for other mining, quarrying		3	2	3	1		...	...	7	...		...	...	...	...
B	Mining and quarrying		108	109	106	105		2934	2395	2435	2359		79	65	68	65
351	Electric power generation, transmission		596	826	1248	1464		6847	6935	6870	6848		200	206	211	220
352	Manufacture of gas		10	10	10	9		136	141	138	137		4	4	4	4
353	Steam and air conditioning supply		42	46	47	53		787	755	772	774		20	19	19	19
D	Electricity, gas, steam & air conditioning		648	882	1305	1526		7770	7831	7781	7759		224	229	234	244
360	Water collection, treatment and supply		70	72	70	71		4167	4007	3819	3908		84	81	79	81
370	Sewerage		30	32	34	40		...	...	311	318		...	...	6	7
381	Waste collection		127	120	119	119		2829	2945	3000	2993		55	58	60	60
382	Waste treatment and disposal		26	32	32	36		276	281	262	266		6	6	5	5
383	Materials recovery		84	102	110	103		1499	1820	1898	1841		30	36	37	34
390	Remediation activities		16	25	27	32		...	...	242	360		...	...	4	7
E	Water supply;sewerage,waste management		353	383	392	401		9303	9547	9532	9687		184	190	191	194

Slovenia

ISIC Revision 4		Output at factor values					Value added at factor values				
			(millions of Euros)					(millions of Euros)			
ISIC	Industry	Note	2010	2011	2012	2013	Note	2010	2011	2012	2013
051	Mining of hard coal		-	-	-	-		-	-	-	-
052	Mining of lignite		...	...	...	...		...	...	...	...
061	Extraction of crude petroleum		-	-	-	-		-	-	-	-
062	Extraction of natural gas		...	...	...	...		...	...	...	...
071	Mining of iron ores		-	-	-	-		-	-	-	-
072	Mining of non-ferrous metal ores		-	-	-	-		-	-	-	-
081	Quarrying of stone, sand and clay		127	122	117	127		46	42	39	40
089	Mining and quarrying n.e.c.		...	...	...	...		...	...	...	...
091	Support activities for petroleum		...	...	...	...		...	...	...	...
099	Support activities for other mining, quarrying		...	...	...	...		...	...	...	...
B	Mining and quarrying		288	261	251	252		142	121	109	104
351	Electric power generation, transmission		2202	2387	2554	2575		645	694	706	740
352	Manufacture of gas		41	42	40	41		17	17	16	15
353	Steam and air conditioning supply		192	201	206	196		49	49	43	49
D	Electricity, gas, steam & air conditioning		2435	2630	2801	2812		711	759	765	804
360	Water collection, treatment and supply		333	321	315	340		130	124	116	125
370	Sewerage		...	...	24	25		...	...	10	11
381	Waste collection		232	232	249	263		88	89	91	98
382	Waste treatment and disposal		34	32	32	31		12	11	10	11
383	Materials recovery		314	417	391	353		75	74	65	53
390	Remediation activities		...	...	14	30		...	...	5	11
E	Water supply;sewerage,waste management		950	1037	1025	1042		319	312	299	308

Spain

ISIC Revision 4		Number of enterprises					Number of employees					Wages and salaries paid to employees				
			(number)					(thousands)					(millions of Euros)			
ISIC	Industry	Note	2010	2011	2012	2013	Note	2010	2011	2012	2013	Note	2010	2011	2012	2013
051	Mining of hard coal		48a/	43a/	32	56		6.1a/	5.4a/	4.8	4.0		213a/	192a/	151	135
052	Mining of lignite		...a/	...a/	-	-		...a/	...a/	-	-		...a/	...a/	-	-
061	Extraction of crude petroleum		4b/	5b/	4	4		0.2b/	0.3b/	0.3	0.3		19b/	21b/	25	32
062	Extraction of natural gas		...b/	...b/	...	...		...b/	...b/	...	...		...b/	...b/	...	...
071	Mining of iron ores		16	18	19	...		0.1	0.1	0.1	...		2	3	2	...
072	Mining of non-ferrous metal ores		34	33	40	12		1.1	1.2	1.4	1.1		38	46	52	41
081	Quarrying of stone, sand and clay		2288	2165	2074	1729		18.5	15.9	13.9	11.8		475	415	362	295
089	Mining and quarrying n.e.c.		150	148	145	102		2.9	2.9	2.8	2.7		98	102	107	104
091	Support activities for petroleum		39c/	32c/	38	48		0.3c/	0.2c/	0.2	0.2		10c/	7c/	8	8
099	Support activities for other mining, quarrying		...c/	...c/	...	4		...c/	...c/	...	-		...c/	...c/	...	1
B	Mining and quarrying		2579	2445	2357	1955		29.2	26.0	23.5	20.3		855	785	708	619
351	Electric power generation, transmission		12817	14711	13722	13650		33.2	36.8	32.9	30.8		1846	2021	1859	1863
352	Manufacture of gas		76	90	77	58		3.1	3.3	4.8	5.0		139	155	320	307
353	Steam and air conditioning supply		206	223	186	160		0.8	0.9	0.8	0.7		14	17	14	12
D	Electricity, gas, steam & air conditioning		13099	15024	13986	13867		37.0	41.0	38.4	36.6		1999	2194	2193	2183
360	Water collection, treatment and supply		2660	2863	2794	2919		40.1	42.6	39.8	39.6		1288	1346	1236	1204
370	Sewerage		497	499	461	480		4.5	4.5	4.2	4.6		110	114	111	121
381	Waste collection		1463	1472	1592	1598		32.1	31.4	33.4	59.0		787	752	809	1458
382	Waste treatment and disposal		415	453	431	356		26.7	27.9	28.3	36.0		688	738	752	943
383	Materials recovery		545	510	568	555		7.9	8.9	8.7	8.7		190	217	209	214
390	Remediation activities		268	228	288	398		1.8	2.2	2.5	1.7		41	51	59	41
E	Water supply;sewerage,waste management		5848	6026	6134	6306		113.1	117.4	116.8	149.5		3103	3218	3176	3981

a/ 051 includes 052.
b/ 061 includes 062.
c/ 091 includes 099.

Spain

ISIC Revision 4		Output (valuation not defined)					Value added at factor values				
			(millions of Euros)					(millions of Euros)			
ISIC	Industry	Note	2010	2011	2012	2013	Note	2010	2011	2012	2013
051	Mining of hard coal		822a/	736a/	552	415		423a/	346a/	175	116
052	Mining of lignite		...a/	...a/	-	-		...a/	...a/	-	-
061	Extraction of crude petroleum		70b/	74b/	77	92		39b/	39b/	43	55
062	Extraction of natural gas		...b/	...b/	...	...		...b/	...b/	...	...
071	Mining of iron ores		9	9	7	...		5	5	2	...
072	Mining of non-ferrous metal ores		462	565	823	411		203	316	476	189
081	Quarrying of stone, sand and clay		2580	2294	1865	1556		979	869	697	606
089	Mining and quarrying n.e.c.		665	795	871	865		268	361	377	361
091	Support activities for petroleum		97c/	103c/	131	246		36c/	41c/	67	153
099	Support activities for other mining, quarrying		...c/	...c/	...	7		...c/	...c/	...	2
B	Mining and quarrying		4704	4576	4341	3614		1953	1976	1848	1496
351	Electric power generation, transmission		37675	54109	51723	63022		14827	19321	19440	17920
352	Manufacture of gas		3698	9013	9777	10621		1554	2487	3286	2471
353	Steam and air conditioning supply		62	98	82	76		29	41	37	36
D	Electricity, gas, steam & air conditioning		41435	63220	61581	73719		16410	21848	22763	20426
360	Water collection, treatment and supply		6708	7245	6931	7213		3664	3570	3289	3527
370	Sewerage		515	451	476	604		226	218	242	261
381	Waste collection		2700	2499	2530	4181		1451	1316	1441	3023
382	Waste treatment and disposal		3119	3222	3326	3960		1372	1741	1645	2064
383	Materials recovery		1843	1863	1676	1751		496	456	426	466
390	Remediation activities		157	199	208	159		69	99	104	73
E	Water supply;sewerage,waste management		15042	15479	15148	17867		7277	7400	7147	9413

a/ 051 includes 052.
b/ 061 includes 062.
c/ 091 includes 099.

Sri Lanka

ISIC Revision 3		Number of establishments					Number of employees					Wages and salaries paid to employees				
			(number)					(number)					(millions of Sri Lankan Rupees)			
ISIC	Industry	Note	2009	2010	2011	2012	Note	2009	2010	2011	2012	Note	2009	2010	2011	2012
10	Mining of coal and lignite; extraction of peat		...	...	...	...		...	...	...	...		...	...	...	...
101	Mining and agglomeration of hard coal		...	...	...	...		...	...	...	...		...	...	...	...
102	Mining and agglomeration of lignite		...	...	...	...		...	...	...	...		...	...	...	...
103	Extraction and agglomeration of peat		...	...	...	...		...	...	...	...		...	...	...	...
11	Extraction of crude oil and natural gas, etc.		...	...	...	...		...	...	...	...		...	...	...	...
111	Extraction of crude oil and natural gas		...	...	...	...		...	...	...	...		...	...	...	...
112	Oil, gas service activities		...	...	...	...		...	...	...	...		...	...	...	...
12	Mining of uranium and thorium ores		...	...	...	...		...	...	...	...		...	...	...	...
120	Mining of uranium and thorium ores		...	...	...	...		...	...	...	...		...	...	...	...
13	Mining of metal ores		...	...	...	...		...	...	...	...		...	...	...	...
131	Iron ores		...	...	...	...		...	...	...	...		...	...	...	...
132	Non-ferrous metal ores,excl.uranium,thorium		...	...	...	...		...	...	...	...		...	...	...	...
14	Other mining and quarrying		...	36	31	26		...	4868	4254	5404		...	683.9	715.5	857.0
141	Quarrying of stone, sand and clay		...	22	19	15		...	1736	1665	1864		...	310.2	297.7	353.9
142	Mining and quarrying n.e.c.		...	14	12	11		...	3132	2589	3540		...	373.8	417.8	503.1
C	Total mining and quarrying	a/	...	36	31	26	a/	...	4868	4254	5404	a/	...	683.9	715.5	857.0
40	Electricity, gas, steam and hot water supply		...	...	...	...		...	...	...	...		...	...	...	...
401	Electricity production, collection, distribution		...	8	5	6		...	15712	16364	16932		...	12344.2	13736.8	11965.3
402	Gas production and distribution		...	...	...	...		...	...	...	...		...	...	...	...
403	Steam and hot water supply		...	...	...	...		...	...	...	...		...	...	...	...
41	Water collection, purification, distribution		...	5	3	4		...	9391	9739	9917		...	2299.6	5210.1	66274.0
410	Water collection, purification, distribution		...	5	3	4		...	9391	9739	9917		...	2299.6	5210.1	66274.0
E	Total electricity, gas and water supply	a/	...	13	8	10	a/	...	25103	26103	26849	a/	...	14643.8	18946.9	78239.3

a/ Sum of available data.

Sri Lanka

ISIC Revision 3		Output at producers' prices					Value added at producers' prices				
			(millions of Sri Lankan Rupees)					(millions of Sri Lankan Rupees)			
ISIC	Industry	Note	2009	2010	2011	2012	Note	2009	2010	2011	2012
10	Mining of coal and lignite; extraction of peat		...	...	...	...		...	...	...	...
101	Mining and agglomeration of hard coal		...	...	...	...		...	...	...	...
102	Mining and agglomeration of lignite		...	...	...	...		...	...	...	...
103	Extraction and agglomeration of peat		...	...	...	...		...	...	...	...
11	Extraction of crude oil and natural gas, etc.		...	...	...	...		...	...	...	...
111	Extraction of crude oil and natural gas		...	...	...	...		...	...	...	...
112	Oil, gas service activities		...	...	...	...		...	...	...	...
12	Mining of uranium and thorium ores		...	...	...	...		...	...	...	...
120	Mining of uranium and thorium ores		...	...	...	...		...	...	...	...
13	Mining of metal ores		...	...	...	...		...	...	...	...
131	Iron ores		...	...	...	...		...	...	...	...
132	Non-ferrous metal ores,excl.uranium,thorium		...	...	...	...		...	...	...	...
14	Other mining and quarrying		...	3612.8	3626.9	5968.6		...	2654.1	2310.9	4506.5
141	Quarrying of stone, sand and clay		...	1729.2	2084.6	4522.3		...	1149.1	1183.3	3523.0
142	Mining and quarrying n.e.c.		...	1883.6	1542.4	1446.2		...	1505.0	1127.6	983.5
C	Total mining and quarrying	a/	...	3612.8	3626.9	5968.6	a/	...	2654.1	2310.9	4506.5
40	Electricity, gas, steam and hot water supply		...	...	...	...		...	...	...	...
401	Electricity production, collection, distribution		...	123213.8	251810.9	356234.4		...	106590.8	135266.8	182069.1
402	Gas production and distribution		...	...	...	...		...	...	...	...
403	Steam and hot water supply		...	...	...	...		...	...	...	...
41	Water collection, purification, distribution		...	12966.8	13504.7	18384.4		...	10262.3	10743.1	14728.9
410	Water collection, purification, distribution		...	12966.8	13504.7	18384.4		...	10262.3	10743.1	14728.9
E	Total electricity, gas and water supply	a/	...	136180.6	265315.6	374618.8	a/	...	116853.0	146009.9	196797.9

a/ Sum of available data.

State of Palestine

ISIC Revision 4		Number of enterprises					Number of persons engaged					Wages and salaries paid to employees				
			(number)					(number)					(thousands of US Dollars)			
ISIC	Industry	Note	2010	2011	2012	2013	Note	2010	2011	2012	2013	Note	2010	2011	2012	2013
05	Mining of coal and lignite		...	...	...	...		...	...	...	...		...	...	...	...
06	Extraction of crude petroleum & natural gas		...	...	...	...		...	...	...	...		...	...	...	...
07	Mining of metal ores		...	...	...	...		...	...	...	...		...	...	...	...
08	Other mining and quarrying		225	247	208	211		1727	1766	1476	1721		8472	9045	9702	10362
09	Mining support service activities		...	...	...	...		...	...	...	...		...	...	...	...
B	Mining and quarrying	a/	225	247	208	211	a/	1727	1766	1476	1721	a/	8472	9045	9702	10362
35	Electricity, gas, steam & air conditioning		7	8	13	13		2111	3352	2522	3301		26141	29225	37863	54219
D	Electricity, gas, steam & air conditioning		7	8	13	13		2111	3352	2522	3301		26141	29225	37863	54219
36	Water collection, treatment and supply		383	419	309b/	313		921	925	912b/	1000		4540	4598	7824b/	8395
37	Sewerage		...	11c/	...b/	...		...	52c/	...b/	...		...	325c/	...b/	...
38	Waste collection,treatment,disposal activities		9	...c/	23d/	24d/		43	...c/	143d/	103d/		241	...c/	468d/	327d/
39	Remediation activities		...	...	...d/	...d/		...	...	...d/	...d/		...	...	...d/	...d/
E	Water supply;sewerage,waste management	a/	392	430	332	337	a/	964	977	1055	1103	a/	4781	4923	8292	8721

a/ Sum of available data.
b/ 36 includes 37.
c/ 37 includes 38.
d/ 38 includes 39.

State of Palestine

ISIC Revision 4		Output (valuation not defined)					Value added (valuation not defined)				
			(thousands of US Dollars)					(thousands of US Dollars)			
ISIC	Industry	Note	2010	2011	2012	2013	Note	2010	2011	2012	2013
05	Mining of coal and lignite		...	...	...	...		...	...	...	...
06	Extraction of crude petroleum & natural gas		...	...	...	...		...	...	...	...
07	Mining of metal ores		...	...	...	...		...	...	...	...
08	Other mining and quarrying		61462	101725	73268	120439		33181	68189	33932	66055
09	Mining support service activities		...	...	...	...		...	...	...	...
B	Mining and quarrying	a/	61462	101725	73268	120439	a/	33181	68189	33932	66055
35	Electricity, gas, steam & air conditioning		339769	409567	319202	600648		167018	220758	177683	194852
D	Electricity, gas, steam & air conditioning		339769	409567	319202	600648		167018	220758	177683	194852
36	Water collection, treatment and supply		37466	11296	36033b/	35262		17162	1816	14743b/	10058
37	Sewerage		...	5590c/	...b/	...		...	616c/	...b/	...
38	Waste collection,treatment,disposal activities		7400	...c/	3627d/	6588d/		1086	...c/	1081d/	2551d/
39	Remediation activities		...	...	...d/	...d/		...	...	...d/	...d/
E	Water supply;sewerage,waste management	a/	44866	...	39660	41850	a/	18248	...	15824	12609

a/ Sum of available data.
b/ 36 includes 37.
c/ 37 includes 38.
d/ 38 includes 39.

Swaziland

ISIC Revision 4		Number of establishments					Number of employees					Wages and salaries paid to employees				
			(number)					(number)					(millions of Swazi Emalangeni)			
ISIC	Industry	Note	2010	2011	2012	2013	Note	2010	2011	2012	2013	Note	2010	2011	2012	2013
05	Mining of coal and lignite		...	1	...	...		...	319	...	...		...	34	...	...
051	Mining of hard coal		...	1	...	...		...	319	...	...		...	34	...	...
052	Mining of lignite		...	...	...	...		...	...	...	...		...	...	...	...
06	Extraction of crude petroleum & natural gas		...	...	...	...		...	...	...	...		...	...	...	...
061	Extraction of crude petroleum		...	...	...	...		...	...	...	...		...	...	...	...
062	Extraction of natural gas		...	...	...	...		...	...	...	...		...	...	...	...
07	Mining of metal ores		...	...	...	...		...	...	...	...		...	...	...	...
071	Mining of iron ores		...	...	...	...		...	...	...	...		...	...	...	...
072	Mining of non-ferrous metal ores		...	...	...	...		...	...	...	...		...	...	...	...
08	Other mining and quarrying		...	6	...	...		...	49	...	...		...	4	...	...
081	Quarrying of stone, sand and clay		...	5	...	...		...	49	...	...		...	4	...	...
089	Mining and quarrying n.e.c.		...	1	...	...		...	...	...	...		...	...	...	...
09	Mining support service activities		...	...	...	...		...	...	...	...		...	...	...	...
091	Support activities for petroleum		...	...	...	...		...	...	...	...		...	...	...	...
099	Support activities for other mining, quarrying		...	...	...	...		...	...	...	...		...	...	...	...
B	Mining and quarrying		...	7a/	...	...		...	368a/	...	...		...	38a/	...	...
35	Electricity, gas, steam & air conditioning		...	10	...	...		...	656	...	...		...	132	...	...
351	Electric power generation, transmission		...	10	...	...		...	645	...	...		...	131	...	...
352	Manufacture of gas		...	...	...	...		...	11	...	...		...	1	...	...
353	Steam and air conditioning supply		...	...	...	...		...	...	...	...		...	...	...	...
D	Electricity, gas, steam & air conditioning		...	10a/	...	...		...	656a/	...	...		...	132a/	...	...
36	Water collection, treatment and supply		...	4	...	...		...	543	...	...		...	70	...	...
37	Sewerage		...	...	...	...		...	...	...	...		...	...	...	...
38	Waste collection,treatment,disposal activities		...	9	...	...		...	120	...	...		...	4	...	...
381	Waste collection		...	...	...	...		...	...	...	...		...	...	...	...
382	Waste treatment and disposal		...	2	...	...		...	54	...	...		...	2	...	...
383	Materials recovery		...	7	...	...		...	66	...	...		...	2	...	...
39	Remediation activities		...	...	...	...		...	...	...	...		...	...	...	...
E	Water supply;sewerage,waste management		...	13a/	...	...		...	663a/	...	...		...	74a/	...	...

a/ Sum of available data.

Swaziland

ISIC Revision 4		Output (valuation not defined) a/					Value added (valuation not defined) b/				
			(millions of Swazi Emalangeni)					(millions of Swazi Emalangeni)			
ISIC	Industry	Note	2010	2011	2012	2013	Note	2010	2011	2012	2013
05	Mining of coal and lignite		...	103	...	...		...	29	...	...
051	Mining of hard coal		...	103	...	...		...	29	...	...
052	Mining of lignite		...	...	...	...		...	...	...	...
06	Extraction of crude petroleum & natural gas		...	...	...	...		...	...	...	...
061	Extraction of crude petroleum		...	...	...	...		...	...	...	...
062	Extraction of natural gas		...	...	...	...		...	...	...	...
07	Mining of metal ores		...	47	...	...		...	43	...	...
071	Mining of iron ores		...	47	...	...		...	43	...	...
072	Mining of non-ferrous metal ores		...	...	...	...		...	...	...	...
08	Other mining and quarrying		...	31	...	...		...	9	...	...
081	Quarrying of stone, sand and clay		...	31	...	...		...	9	...	...
089	Mining and quarrying n.e.c.		...	...	...	...		...	...	...	...
09	Mining support service activities		...	...	...	...		...	...	...	...
091	Support activities for petroleum		...	...	...	...		...	...	...	...
099	Support activities for other mining, quarrying		...	...	...	...		...	...	...	...
B	Mining and quarrying		...	181c/	...	...		...	82c/	...	...
35	Electricity, gas, steam & air conditioning		...	1050	...	...		...	504	...	...
351	Electric power generation, transmission		...	1043	...	...		...	500	...	...
352	Manufacture of gas		...	7	...	...		...	4	...	...
353	Steam and air conditioning supply		...	...	...	...		...	...	...	...
D	Electricity, gas, steam & air conditioning		...	1050c/	...	...		...	504c/	...	...
36	Water collection, treatment and supply		...	233	...	...		...	161	...	...
37	Sewerage		...	...	...	...		...	...	...	...
38	Waste collection,treatment,disposal activities		...	24	...	...		...	19	...	...
381	Waste collection		...	...	...	...		...	...	...	...
382	Waste treatment and disposal		...	15	...	...		...	12	...	...
383	Materials recovery		...	9	...	...		...	6	...	...
39	Remediation activities		...	...	...	...		...	...	...	...
E	Water supply;sewerage,waste management		...	257c/	...	...		...	180c/	...	...

a/ Output is at basic prices.
b/ Value added is at basic prices.
c/ Sum of available data.

Sweden

ISIC Revision 4		Number of enterprises					Number of employees					Wages and salaries paid to employees				
		Note	(number)				Note	(number)				Note	(millions of Swedish Kronor)			
ISIC	Industry		2010	2011	2012	2013		2010	2011	2012	2013		2010	2011	2012	2013
051	Mining of hard coal		...	...	...	...		...	...	...	...		...	...	...	...
052	Mining of lignite		...	...	...	...		...	...	...	...		...	...	...	...
061	Extraction of crude petroleum		...	...	...	...		...	...	...	...		...	...	...	...
062	Extraction of natural gas		...	...	...	...		...	...	...	...		...	...	...	...
071	Mining of iron ores		13	12	11	13		3182	3434	3641	3975		1370	1718	1760	1899
072	Mining of non-ferrous metal ores		26	27	24	26		...	...	...	...		...	...	...	...
081	Quarrying of stone, sand and clay		497	500	495	479		2665	2532	2567	2484		932	848	866	808
089	Mining and quarrying n.e.c.		105	103	102	97		259	306	...	239		74	85	...	79
091	Support activities for petroleum		45	47	50	58		...	...	...	49		...	...	...	26
099	Support activities for other mining, quarrying		64	62	71	76		107	129	...	...		58	79	...	...
B	Mining and quarrying		750	751	753	749		8337	8748	9200	9478		3382	3772	3923	4065
351	Electric power generation, transmission		1550	1709	1867	1957		20078	19710	19712	19772		8645	8965	9110	9398
352	Manufacture of gas		45	57	67	76		262	267	272	285		125	140	144	153
353	Steam and air conditioning supply		233	247	248	254		8067	8083	8287	8331		3243	3401	3564	3675
D	Electricity, gas, steam & air conditioning		1828	2013	2182	2287		28407	28060	28271	28388		12013	12505	12818	13226
360	Water collection, treatment and supply		196	212	218	226		992	1196	1302	1448		362	464	496	581
370	Sewerage		176	188	200	212		1705	1674	1905	2105		532	569	672	754
381	Waste collection		267	277	280	288		7482	8241	7941	7921		2371	2696	2598	2694
382	Waste treatment and disposal		120	125	134	137		1019	1384	1516	1425		329	548	606	607
383	Materials recovery		361	389	412	426		3574	3905	4349	4450		1140	1289	1554	1607
390	Remediation activities		106	113	117	116		221	287	324	352		51	81	84	97
E	Water supply;sewerage,waste management		1226	1304	1361	1405		14994	16686	17339	17701		4785	5647	6011	6340

Sweden

ISIC Revision 4		Output (valuation not defined)					Value added at factor values				
			(millions of Swedish Kronor)					(millions of Swedish Kronor)			
ISIC	Industry	Note	2010	2011	2012	2013	Note	2010	2011	2012	2013
051	Mining of hard coal		...	...	...	...		...	...	...	...
052	Mining of lignite		...	...	...	...		...	...	...	...
061	Extraction of crude petroleum		...	...	...	...		...	...	...	...
062	Extraction of natural gas		...	...	...	...		...	...	...	...
071	Mining of iron ores		25988	28271	25440	23605		15135	17593	14052	10653
072	Mining of non-ferrous metal ores		...	...	...	...		...	...	...	...
081	Quarrying of stone, sand and clay		7402	7877	8222	7435		1922	2319	2327	2158
089	Mining and quarrying n.e.c.		487	521	...	493		184	211	...	189
091	Support activities for petroleum		...	...	...	1032		...	...	...	606
099	Support activities for other mining, quarrying		355	1183	...	...		-98	-87	...	...
B	Mining and quarrying		45392	49696	45620	41027		24941	27600	23380	18115
351	Electric power generation, transmission		124557	134992	133880	128449		67745	64805	64301	57912
352	Manufacture of gas		1542	1965	2079	2039		849	722	908	805
353	Steam and air conditioning supply		40420	37187	38916	38068		14508	14450	15823	16373
D	Electricity, gas, steam & air conditioning		166516	174144	174875	168556		83095	79989	81032	75092
360	Water collection, treatment and supply		3641	4307	4960	5202		1348	1559	1816	1886
370	Sewerage		3494	3627	3968	4404		1398	1537	1754	1972
381	Waste collection		13049	14373	13589	13903		5265	5813	5496	6079
382	Waste treatment and disposal		2728	4139	4621	4109		927	1429	1769	2032
383	Materials recovery		16873	20518	21166	19958		3007	3531	3705	10082
390	Remediation activities		265	459	412	457		94	169	152	272
E	Water supply;sewerage,waste management		40052	47424	48714	48034		12040	14037	14691	22323

Switzerland

ISIC Revision 4		Number of enterprises					Number of persons engaged					Wages and salaries paid to employees			
		Note	(number)				Note	(number)				Note	(millions of Swiss Francs)			
ISIC	Industry		2010	2011	2012	2013		2010	2011	2012	2013		2010	2011	2012	2013
051	Mining of hard coal		...	...	...	...		...	...	...	...		...	...	...	...
052	Mining of lignite		...	...	...	...		...	...	...	...		...	...	...	...
061	Extraction of crude petroleum		...	1	1	...		...	28	...	...		...	...	...	...
062	Extraction of natural gas		...	-	-	...		...	-	...	...		...	...	...	...
071	Mining of iron ores		...	...	...	...		...	...	...	...		...	...	...	...
072	Mining of non-ferrous metal ores		...	...	...	...		...	...	...	...		...	...	...	...
081	Quarrying of stone, sand and clay		186	182	180	202		3853	3822	3696	3946		316	343a/	317	...
089	Mining and quarrying n.e.c.		9	7	7	10		329	253	368	349		...	...a/	32	...
091	Support activities for petroleum		4	5	5	4		67	74	71	61		...	...	12	...
099	Support activities for other mining, quarrying		-	-	-	-		...	...	-	-		...	...	-	...
B	Mining and quarrying		199	195	193	216		4250	4177	4134	4355		367	367	361	368
351	Electric power generation, transmission		316	308	311	406		26166	26980	27048	27724		2648	2663	2653	2720
352	Manufacture of gas		21	22	25	33		922	956	937	970		97	100	...	...
353	Steam and air conditioning supply		9	9	10	20		9	...	498	46		1	...	...	...
D	Electricity, gas, steam & air conditioning		346	339	346	459		27097	27936	28484	28740		2745	2763	2798	2827
360	Water collection, treatment and supply		55	50	49	95		1114	1017	1055	1226		90	91	89	89
370	Sewerage		172	169	168	170		2426	2605	2769	2645		202	213	226	224
381	Waste collection		132	127	133	129		2193	2664	2852	2216		165	203	202	150
382	Waste treatment and disposal		84	90	98	111		1988	1806	1570	2424		162	150	130	190
383	Materials recovery		187	184	186	189		4709	3955	4750	4799		332	301	360	360
390	Remediation activities		10	16	19	18		81	114	232	231		5	7	13	14
E	Water supply;sewerage,waste management		640	636	653	712		12510	12161	13228	13540		956	965	1020	1026

a/ 081 includes 089.

Switzerland

ISIC Revision 4		Output (valuation not defined)					Value added at factor values				
		Note	(millions of Swiss Francs)				Note	(millions of Swiss Francs)			
ISIC	Industry		2010	2011	2012	2013		2010	2011	2012	2013
051	Mining of hard coal		...	...	...	...		...	...	...	...
052	Mining of lignite		...	...	...	...		...	...	...	...
061	Extraction of crude petroleum		...	...	...	...		...	...	...	...
062	Extraction of natural gas		...	...	...	...		...	...	...	...
071	Mining of iron ores		...	...	...	...		...	...	...	...
072	Mining of non-ferrous metal ores		...	...	...	...		...	...	...	...
081	Quarrying of stone, sand and clay		1775	2042a/	1984	...		739	818a/	758	...
089	Mining and quarrying n.e.c.		...	...a/	169	...		...	...a/	81	...
091	Support activities for petroleum		...	...	53	...		...	...	18	...
099	Support activities for other mining, quarrying		...	...	-	...		...	...	-	...
B	Mining and quarrying		1995	2090	2206	2216		874	850	857	861
351	Electric power generation, transmission		39118	34592	40806	39953		9046	8350	9201	8872
352	Manufacture of gas		3150	3100	...	...		381	350	...	...
353	Steam and air conditioning supply		...	...	...	...		2	-	...	...
D	Electricity, gas, steam & air conditioning		42273	37692	44285	52664		9429	8699	9622	9364
360	Water collection, treatment and supply		653	639	698	646		260	242	268	249
370	Sewerage		694	756	828	813		382	396	435	422
381	Waste collection		748	960	735	775		313	452	392	291
382	Waste treatment and disposal		1167	1015	841	1392		610	456	371	509
383	Materials recovery		2129	2635	2664	2411		601	579	659	623
390	Remediation activities		16	16	31	32		5	6	18	14
E	Water supply;sewerage,waste management		5407	6022	5798	6070		2170	2130	2143	2108

a/ 081 includes 089.

Tajikistan

ISIC Revision 3		Number of establishments					Number of employees					Wages and salaries				
		Note	(number)				Note	(number)				Note	(Tajikistan Somoni)			
ISIC	Industry		2010	2011	2012	2013		2010	2011	2012	2013		2010	2011	2012	2013
10	Mining of coal and lignite; extraction of peat		...	...	...	...		...	...	...	...		...	...	...	...
101	Mining and agglomeration of hard coal		...	14	12	14		...	370	375	393		...	...	...	...
102	Mining and agglomeration of lignite		...	2	2	2		...	295	307	335		...	...	...	...
103	Extraction and agglomeration of peat		...	...	...	...		...	...	...	...		...	...	...	...
11	Extraction of crude oil and natural gas, etc.		...	...	9	4		...	...	926	189		...	...	...	...
111	Extraction of crude oil and natural gas		...	9	8	2		...	954	786	13		...	...	...	...
112	Oil, gas service activities		...	...	1	2		...	...	140	176		...	...	...	...
12	Mining of uranium and thorium ores		...	...	...	...		...	...	...	...		...	...	...	...
120	Mining of uranium and thorium ores		...	...	...	...		...	...	...	...		...	...	...	...
13	Mining of metal ores		...	...	...	...		...	...	...	...		...	...	...	...
131	Iron ores		...	...	...	...		...	...	...	...		...	...	...	...
132	Non-ferrous metal ores,excl.uranium,thorium		...	17	15	15		...	7629	6377	4194		...	...	...	...
14	Other mining and quarrying		...	54	71	92		...	1180	1084	1500		...	...	...	...
141	Quarrying of stone, sand and clay		...	43	59	79		...	841	795	1169		...	...	...	...
142	Mining and quarrying n.e.c.		...	11	12	13		...	339	289	331		...	...	...	...
C	Total mining and quarrying		...	96	109	136		...	10428	9069	7329		...	...	...	...
40	Electricity, gas, steam and hot water supply		...	24	24	23		...	14605	15434	13454		...	...	...	...
401	Electricity production, collection, distribution		...	18	19	18		...	13138	14666	12686		...	...	...	...
402	Gas production and distribution		...	5	4	4		...	685	482	482		...	...	...	...
403	Steam and hot water supply		...	1	1	1		...	782	286	286		...	...	...	...
41	Water collection, purification, distribution		...	5	5	5		...	2607	3083	3080		...	...	...	...
410	Water collection, purification, distribution		...	5	5	5		...	2607	3083	3080		...	...	...	...
E	Total electricity, gas and water supply		...	29	29	28		...	17212	18517	16534		...	...	...	...

Tajikistan

ISIC Revision 3		Output (valuation not defined)					Value added				
		Note	(thousands of Tajikistan Somoni)				Note	(thousands of Tajikistan Somoni)			
ISIC	Industry		2010	2011	2012	2013		2010	2011	2012	2013
10	Mining of coal and lignite; extraction of peat		...	...	...	...		...	...	...	...
101	Mining and agglomeration of hard coal		...	21379	47256	75524		...	...	...	...
102	Mining and agglomeration of lignite		...	4898	10326	12198		...	...	...	...
103	Extraction and agglomeration of peat		...	...	...	...		...	...	...	...
11	Extraction of crude oil and natural gas, etc.		...	...	54479	172806		...	...	...	...
111	Extraction of crude oil and natural gas		...	48242	49479	840		...	...	...	...
112	Oil, gas service activities		...	...	5000	171966		...	...	...	...
12	Mining of uranium and thorium ores		...	...	...	...		...	...	...	...
120	Mining of uranium and thorium ores		...	...	...	...		...	...	...	...
13	Mining of metal ores		...	...	...	...		...	...	...	...
131	Iron ores		...	...	...	...		...	...	...	...
132	Non-ferrous metal ores,excl.uranium,thorium		...	883925	1077834	751968		...	...	...	...
14	Other mining and quarrying		...	16229	36257	51549		...	...	...	...
141	Quarrying of stone, sand and clay		...	12064	31748	45776		...	...	...	...
142	Mining and quarrying n.e.c.		...	4165	4509	5773		...	...	...	...
C	Total mining and quarrying		...	974673	1226153	1116995		...	...	...	...
40	Electricity, gas, steam and hot water supply		...	1423870	1699794	1866042		...	...	...	...
401	Electricity production, collection, distribution		...	1416512	1695393	1865062		...	...	...	...
402	Gas production and distribution		...	7084	4103	591		...	...	...	...
403	Steam and hot water supply		...	274	298	389		...	...	...	...
41	Water collection, purification, distribution		...	34711	40788	53242		...	...	...	...
410	Water collection, purification, distribution		...	34711	40788	53242		...	...	...	...
E	Total electricity, gas and water supply		...	1458581	1740583	1919283		...	...	...	...

Turkey

ISIC Revision 4		Number of enterprises					Number of employees					Wages and salaries paid to employees				
			(number)					(number)					(millions of Turkish Liras)			
ISIC	Industry	Note	2009	2010	2011	2012	Note	2009	2010	2011	2012	Note	2009	2010	2011	2012
051	Mining of hard coal		83	93	66	71		17297	16967	17308	18224		438.2	479.8	548.6	566.1
052	Mining of lignite		218	291	233	242		32519	33740	32967	32527		695.4	701.4	783.3	779.5
061	Extraction of crude petroleum		6	6	6	8	a/	5334	5541	5673	5446	a/	307.3	329.5	384.0	416.8
062	Extraction of natural gas		3	4	6	6	a/	...	...	...	...	a/	...	...	...	...
071	Mining of iron ores		25	26	27	34		1688	1776	2198	2322		36.0	43.3	56.5	68.2
072	Mining of non-ferrous metal ores		148	194	266	339		9846	14880	19870	19335		183.8	282.2	427.5	496.2
081	Quarrying of stone, sand and clay		1506	1671	2125	2154		22803	27248	30951	34523		240.4	340.3	389.0	489.2
089	Mining and quarrying n.e.c.		392	351	430	472		8732	9222	10176	13199		178.8	204.0	250.3	321.9
091	Support activities for petroleum		12	19	26	28		931	1283	1799	2014		31.1	57.4	91.6	97.3b/
099	Support activities for other mining, quarrying		-	8	10	12		-	122	310	161		-	2.0	3.0	...b/
B	Mining and quarrying		2393	2663	3195	3366		99150	110779	121252	127751		2110.9	2439.9	2933.7	3235.1
351	Electric power generation, transmission		200	264	414	546		56483	56257	59069	63443		2054.3	2324.6	2440.3	2655.5
352	Manufacture of gas		51	62	66	71		5044	5332	5768	6045		193.6	227.8	284.0	284.8
353	Steam and air conditioning supply		28	31	40	54		474	568	560	604		7.4	9.1	10.2	12.8
D	Electricity, gas, steam & air conditioning		279	357	520	671		62001	62157	65397	70092		2255.3	2561.4	2734.6	2953.1
360	Water collection, treatment and supply		2639	2493	2665	2817		32545	32365	33421	33204		1021.2	1090.0	1237.4	1363.2
370	Sewerage		78	89	124	154		690	1008	1425	1324		9.4	19.9	26.2	32.8
381	Waste collection		116	153	206	279		33021	40020	46329	50137		452.3	543.7	708.7	848.3
382	Waste treatment and disposal		12	16	25	35		576	635	806	1287		18.4	28.6	22.3	35.6
383	Materials recovery		90	154	299	407		1800	3290	6319	7453		20.3	41.2	81.0	110.3
390	Remediation activities		7	9	12	14		139	272	116	173		2.9	4.7	2.1	3.5
E	Water supply;sewerage,waste management		2942	2914	3331	3706		68771	77590	88416	93578		1524.4	1728.1	2077.7	2393.9

a/ 061 includes 062.
b/ 091 includes 099.

Turkey

ISIC Revision 4		Output (valuation not defined)					Value added at factor values				
			(millions of Turkish Liras)					(millions of Turkish Liras)			
ISIC	Industry	Note	2009	2010	2011	2012	Note	2009	2010	2011	2012
051	Mining of hard coal		559.7	516.4	719.8	694.0		221.3	241.1	311.4	281.5
052	Mining of lignite		4205.9	4058.4	4278.4	4853.9		1854.6	1705.2	1993.5	2108.5
061	Extraction of crude petroleum	a/	2515.8	2567.5	4421.2	3343.5	a/	797.3	1197.9	1445.2	1711.6
062	Extraction of natural gas	a/	...	...	...	...	a/	...	...	...	...
071	Mining of iron ores		449.8	586.9	931.1	1022.8		163.4	272.0	435.7	486.7
072	Mining of non-ferrous metal ores		1948.5	2832.0	4985.8	6281.5		1113.4	1636.7	2936.2	3616.2
081	Quarrying of stone, sand and clay		2493.5	3571.6	5054.9	5308.4		579.0	805.3	1078.1	1204.2
089	Mining and quarrying n.e.c.		1365.1	1993.9	3245.2	3395.4		681.3	965.9	1530.2	1638.2
091	Support activities for petroleum		385.8	2063.6	1661.4	595.0		108.4	535.1	376.5	266.7
099	Support activities for other mining, quarrying		-	9.8	17.6	12.5		-	3.5	6.1	0.9
B	Mining and quarrying		13924.0	18200.2	25315.4	25507.1		5518.7	7362.7	10113.0	11314.5
351	Electric power generation, transmission		68809.8	74722.9	85630.7	109237.7		10667.6	14453.3	11992.3	14062.1
352	Manufacture of gas		7046.8	6372.1	9694.9	11896.7		663.3	664.9	942.2	950.0
353	Steam and air conditioning supply		48.0	60.3	63.4	92.7		20.4	24.4	24.5	42.7
D	Electricity, gas, steam & air conditioning		75904.5	81155.4	95389.1	121227.1		11351.2	15142.6	12959.0	15054.8
360	Water collection, treatment and supply		5517.6	6428.6	7698.8	9606.7		2792.5	3276.5	4373.0	5488.0
370	Sewerage		51.9	130.7	153.0	323.8		24.0	44.1	40.4	...
381	Waste collection		1473.6	1633.4	2109.7	2899.6		744.5	813.6	1079.0	1383.8
382	Waste treatment and disposal		67.0	68.2	139.4	236.4		31.1	26.4	44.8	91.1
383	Materials recovery		491.6	1046.7	2887.9	3948.0		19.2	102.5	222.3	240.6
390	Remediation activities		9.3	19.0	11.4	17.7		4.2	6.8	3.0	...
E	Water supply;sewerage,waste management		7611.0	9326.6	13000.2	17032.1		3615.5	4269.9	5762.4	7233.1

a/ 061 includes 062.

Ukraine

ISIC Revision 4		Number of establishments					Number of employees					Wages and salaries paid to employees				
			(number)					(number)					(millions of Ukrainian Hryvnias)			
ISIC	Industry	Note	2010	2011	2012	2013	Note	2010	2011	2012	2013	Note	2010	2011	2012	2013
051	Mining of hard coal		274	313	365	...a/		313306	...a/	...a/	...a/		13018	...a/	...a/	...a/
052	Mining of lignite		6	1	2	...a/		327	...a/	...a/	...a/		5	...a/	...a/	...a/
061	Extraction of crude petroleum		37	34	39	41		30314	29317	28253	27246		863	841	1376	1329
062	Extraction of natural gas		39	45	62	78		26337	26961	27445	28371		780	958	1984	2053
071	Mining of iron ores		17	18	29	31		55282	45794	53079	52916		2260	2325	3008	3163
072	Mining of non-ferrous metal ores		18	18	19	28		30584	30943	19485	18536		989	1187	863	884
081	Quarrying of stone, sand and clay		948	919	918	1097		41890	42605	38716	36358		1020	1269	1311	1283
089	Mining and quarrying n.e.c.		82	76	80	112		5271	5890	5719	5179		197	252	275	288
091	Support activities for petroleum		50	51	56	82		1526	1517	1445	1911		92	77	101	121
099	Support activities for other mining, quarrying		158	85	52	50		1741	13633	2578	2639		55	692	112	109
B	Mining and quarrying		1629	1560	1622	1918		506578	487377	482587	473841		19279	22972	27827	29207
351	Electric power generation, transmission		425	463	538	819		244781	227512	224247	222180		9962	11097	13035	14374
352	Manufacture of gas		229	251	216	235		86532	88614	88879	87718		2330	2876	3015	3055
353	Steam and air conditioning supply		602	599	580	669		100496	110742	108166	106823		2749	3850	4288	4595
D	Electricity, gas, steam & air conditioning		1256	1313	1334	1723		431809	426868	421292	416721		15041	17823	20338	22024
360	Water collection, treatment and supply		1491	1531	1651	1786		106092	103933	102914	100137		2506	2727	3048	3182
370	Sewerage		228	222	211	257		16105	12602	14129	19738		380	314	403	621
381	Waste collection		978	1033	912	1093		33734	33398	28038	25956		678	806	786	753
382	Waste treatment and disposal		57	101	102	132		1375	2096	1873	1763		33	55	52	60
383	Materials recovery		934	970	735	784		13010	13075	12506	10711		239	330	308	350
390	Remediation activities		36	36	32	38		993	989	706	383		19	23	20	10
E	Water supply;sewerage,waste management		3724	3893	3643	4090		171309	166093	160166	158688		3855	4255	4617	4976

a/ Data suppressed due to confidentiality rules.

Ukraine

ISIC Revision 4			Output at factor values					Value added at factor values			
			(millions of Ukrainian Hryvnias)					(millions of Ukrainian Hryvnias)			
ISIC	Industry	Note	2010	2011	2012	2013	Note	2010	2011	2012	2013
051	Mining of hard coal	a/	...	...	...	...	a/	...	...	...	...
052	Mining of lignite	a/	...	...	...	...	a/	...	...	...	...
061	Extraction of crude petroleum		...	...	14776	14578		...	...	11317	6410
062	Extraction of natural gas		...	...	20080	21547		...	...	11174	12255
071	Mining of iron ores		...	...	51281	58083		...	...	25683	31720
072	Mining of non-ferrous metal ores		...	...	4316	3616		...	...	2576	1931
081	Quarrying of stone, sand and clay		...	...	11290	10570		...	...	4610	4414
089	Mining and quarrying n.e.c.		...	...	1561	1447		...	...	928	848
091	Support activities for petroleum		...	...	810	834		...	...	356	97
099	Support activities for other mining, quarrying		...	...	724	790		...	...	87	118
B	Mining and quarrying		...	...	157921	170060		...	...	94292	93376
351	Electric power generation, transmission		...	...	107387	107321		...	...	49264	47020
352	Manufacture of gas		...	...	15962	17635		...	...	9452	12106
353	Steam and air conditioning supply		...	...	27794	24234		...	...	7170	408
D	Electricity, gas, steam & air conditioning		...	...	151143	149190		...	...	65887	59534
360	Water collection, treatment and supply		...	...	9885	9055		...	...	7294	4887
370	Sewerage		...	...	1121	1684		...	...	645	893
381	Waste collection		...	...	2946	3079		...	...	1574	1490
382	Waste treatment and disposal		...	...	223	256		...	...	107	112
383	Materials recovery		...	...	7196	6977		...	...	1216	1234
390	Remediation activities		...	...	106	54		...	...	58	19
E	Water supply;sewerage,waste management		...	...	21477	21106		...	...	10894	8636

a/ Data suppressed due to confidentiality rules.

United Kingdom

ISIC Revision 4		Number of enterprises					Number of employees					Wages and salaries paid to employees				
			(number)					(number)					(millions of British Pounds)			
ISIC	Industry	Note	2010	2011	2012	2013	Note	2010	2011	2012	2013	Note	2010	2011	2012	2013
051	Mining of hard coal		23	21	...	20		6001	6504	7962	5421		274	252	...	214
052	Mining of lignite		-	-	-	-		-	-	-	-		-	-	-	-
061	Extraction of crude petroleum		83	127a/	98	92		14703	12605	14239	13694	a/	1496	1653	1876	2070
062	Extraction of natural gas		15	...a/	53	66		470	420	860	666	a/	...	...	...	...
071	Mining of iron ores		...	-	-	7b/		1	-	-	...		-	-	-	-b/
072	Mining of non-ferrous metal ores		...	-	...	...b/		-	-	...	...		-	-	...	...b/
081	Quarrying of stone, sand and clay		513	490	458	434		14234	16354	22922	17376		392	415	480	642
089	Mining and quarrying n.e.c.		196	228	245	275		2488	...	3861	2136		101	74	81	93
091	Support activities for petroleum		270	264	251	286		22784	...	...	...		1245	1347	1424	1575
099	Support activities for other mining, quarrying		53	76	127	123		141	...	...	...		-	3	1	10
B	Mining and quarrying		1153	1206	1263	1303		60822	62264	74827	65431		3507	3746	4164	4604
351	Electric power generation, transmission		563	1118	1710	2459		83078	81568	80214	86956		2770	2919	3125	3074
352	Manufacture of gas		71	...	98	...		40321	41704	...	...		1472	...	1628	...
353	Steam and air conditioning supply		17	...	21	...		258	162	...	...		...	...	-	...
D	Electricity, gas, steam & air conditioning		651	1220	1829	2577		123658	123433	120087	129454		4242	4432	4754	4708
360	Water collection, treatment and supply		118	119	120	106		...	...	...	40143		1099	1102	1164	1295
370	Sewerage		937	896	899	924		11009	10918	14709	18031		361	323	322	347
381	Waste collection		2054	2187	2358	2377		48023	56000	51387	47907		1033	1069	1126	1197
382	Waste treatment and disposal		705	803	946	988		...	26716	24483	25015		459	590	643	652
383	Materials recovery		1840	1952	1884	1982		...	26698	27335	28490		580	603	629	687
390	Remediation activities		346	458	603	696		...	...	...	4030		30	68	57	57
E	Water supply;sewerage,waste management		6000	6415	6810	7073		149599	156082	159688	163617		3563	3755	3942	4235

a/ 061 includes 062.
b/ 071 includes 072.

United Kingdom

ISIC Revision 4			Output (valuation not defined)					Value added at factor values			
			(millions of British Pounds)					(millions of British Pounds)			
ISIC	Industry	Note	2010	2011	2012	2013	Note	2010	2011	2012	2013
051	Mining of hard coal		967	1145	...	...		365	425	...	...
052	Mining of lignite		-	-	-	-		-	-	-	-
061	Extraction of crude petroleum	a/	35437	41933	36183	36841	a/	22556	25368	19650	18451
062	Extraction of natural gas	a/	...	...	...	...	a/	...	...	...	...
071	Mining of iron ores		-	-	-	...		-	-	-	...
072	Mining of non-ferrous metal ores		-	-	...	...		-	-	...	...
081	Quarrying of stone, sand and clay		2838	3035	3502	5274		746	811	937	1556
089	Mining and quarrying n.e.c.		443	486	475	550		130	250	237	228
091	Support activities for petroleum		6833	6026	6380	7532		3218	2547	3052	3134
099	Support activities for other mining, quarrying		2	22	10	13		1	11	8	-18
B	Mining and quarrying		46521	52652	47801	51060		27015	29414	24253	23655
351	Electric power generation, transmission		31576	34505	37643	38197		14705	16740	17002	16267
352	Manufacture of gas		...	...	10422	...		...	...	6173	...
353	Steam and air conditioning supply		...	...	2	...		...	...	1	...
D	Electricity, gas, steam & air conditioning		41619	44206	48067	49762		20731	22451	23177	22342
360	Water collection, treatment and supply		10068	10398	10914	12092		7083	7585	7939	8997
370	Sewerage		2578	2517	2778	2816		1788	1831	1828	1570
381	Waste collection		5606	5486	6440	6741		2152	2095	2675	2571
382	Waste treatment and disposal		2690	3795	3165	3933		973	1500	1136	1326
383	Materials recovery		7404	9019	7931	6965		1850	2138	1958	1398
390	Remediation activities		185	216	304	376		94	109	196	164
E	Water supply;sewerage,waste management		28531	31432	31531	32923		13940	15259	15733	16026

a/ 061 includes 062.

United States of America

ISIC Revision 4		Number of enterprises					Number of employees					Wages and salaries paid to employees				
			(number)					(number)					(millions of US Dollars)			
ISIC	Industry	Note	2010	2011	2012	2013	Note	2010	2011	2012	2013	Note	2010	2011	2012	2013
051	Mining of hard coal		2143	2203	2169	2155		20761	23330	28172	27445		2263	2761	3578	3403
052	Mining of lignite		544	539	513	473		90113	89724	97057	91642		7391	7719	8255	7791
061	Extraction of crude petroleum		2117	2133	2131	2118		26774	34942	24793	38469		3374	5211	2796	5306
062	Extraction of natural gas		2181	2233	2263	2223		49697	54663	77508	77045		6299	6843	12301	11146
071	Mining of iron ores		17	15	16	11		...	...	...	...		...	...	...	...
072	Mining of non-ferrous metal ores		236	264	297	282		29970	36614	41732	44653		2759	3502	3857	4429
081	Quarrying of stone, sand and clay		2540	2487	2378	2296		57431	54801	55505	58879		3347	3331	3272	3635
089	Mining and quarrying n.e.c.		523	526	545	532		17457	22506	18615	16872		1078	1410	1186	1046
091	Support activities for petroleum		9551	10067	10852	10945		307726	355569	438094	408813		23828	29401	36072	34700
099	Support activities for other mining, quarrying		316	333	321	309		5897	7668	8411	7320		423	562	597	485
B	Mining and quarrying		20168	20800	21485	21344		610735	685491	795928	776603		51193	61257	72529	72479
351	Electric power generation, transmission		1663	1689	1741	1723		557240	536017	537250	532284		50287	50596	51460	52228
352	Manufacture of gas		373	383	414	391		72227	76270	74335	73592		5491	6035	6393	6297
353	Steam and air conditioning supply		372	351	342	333		5564	6230	6708	6405		240	247	280	266
D	Electricity, gas, steam & air conditioning		2408	2423	2497	2447		635031	618517	618293	612281		56018	56878	58133	58790
360	Water collection, treatment and supply		1637	1631	1618	1627		12037	17427	11554	18169		547	946	542	1096
370	Sewerage		4594	4538	4633	4666		34866	35123	35656	37131		1445	1482	1521	1653
381	Waste collection		7560	7508	7681	7716		205709	199516	195479	206623		10047	10106	9857	10569
382	Waste treatment and disposal		1330	1238	1241	1127		41741	44400	53259	52418		2413	2484	3395	3198
383	Materials recovery		1010	1005	1131	1158		14785	16391	17880	20002		555	627	658	869
390	Remediation activities		3078	3082	3192	3237		62237	62518	63551	65373		3358	3296	3490	3642
E	Water supply;sewerage,waste management		19209	19002	19496	19531		371375	375375	377379	399716		18365	18941	19464	21027

Viet Nam

ISIC Revision 4		Number of enterprises					Number of persons engaged					Wages and salaries paid to employees				
			(number)					(number)					(billions of Vietnamese Dongs)			
ISIC	Industry	Note	2010	2011	2012	2013	Note	2010	2011	2012	2013	Note	2010	2011	2012	2013
05	Mining of coal and lignite		115	70	84	82		100816	95156	104903	104337		8675	9748	10386	9938
051	Mining of hard coal		102	59	72	70		100434	94871	104614	104055		8666	9737	10372	9927
052	Mining of lignite		13	11	12	12		382	285	289	282		9	11	14	12
06	Extraction of crude petroleum & natural gas		8	7	9	8		7643	8200	9770	10779		3160	1546	2465	4632
061	Extraction of crude petroleum		6	5	5	6		7498	8056	8536	9538		3143	1525	2121	4229
062	Extraction of natural gas		2	2	4	2		145	144	1234	1241		17	21	343	403
07	Mining of metal ores		171	271	306	295		14702	19062	18812	18949		507	1027	1159	1086
071	Mining of iron ores		73	115	125	125		3681	5664	5676	5968		155	307	331	314
072	Mining of non-ferrous metal ores		98	156	181	170		11021	13398	13136	12981		352	720	828	773
08	Other mining and quarrying		1869	2088	2123	2060		64848	63898	58104	53931		2795	2946	3235	3124
081	Quarrying of stone, sand and clay		1761	1971	2000	1952		57052	55464	50823	46610		2320	2393	2466	2488
089	Mining and quarrying n.e.c.		108	117	123	108		7796	8434	7281	7321		475	553	769	636
09	Mining support service activities		30	70	80	112		2621	7897	8705	8958		458	1181	1253	1944
091	Support activities for petroleum		9	24	29	48		1233	1894	2499	3296		329	605	964	1410
099	Support activities for other mining, quarrying		21	46	51	64		1388	6003	6206	5662		128	576	289	534
B	Mining and quarrying		2193	2506	2602	2557		190630	194213	200294	196954		15595	16448	18498	20725
35	Electricity, gas, steam & air conditioning		910	1050	1086	1083		168319	199223	203755	221939		12234	16131	18394	27644
351	Electric power generation, transmission		187	265	321	321		159877	190266	196111	214017		11276	15558	18100	27302
352	Manufacture of gas		11	12	15	9		1541	1401	365	432		36	353	20	33
353	Steam and air conditioning supply		712	773	750	753		6901	7556	7279	7490		922	220	274	309
D	Electricity, gas, steam & air conditioning		910	1050	1086	1083		168319	199223	203755	221939		12234	16131	18394	27644
36	Water collection, treatment and supply		286	313	364	381		29434	31195	32586	33950		1995	2211	2712	3233
37	Sewerage		153	127	189	170		6940	4923	7638	6696		388	435	653	585
38	Waste collection,treatment,disposal activities		382	473	547	547		31406	45867	46809	51480		1243	2611	3262	3765
381	Waste collection		248	324	346	354		28128	41813	42064	45550		1118	2411	2998	3406
382	Waste treatment and disposal		46	60	93	90		1776	2540	3088	4186		83	158	209	277
383	Materials recovery		88	89	108	103		1502	1514	1657	1744		42	43	55	82
39	Remediation activities		29	18	33	27		948	421	361	264		54	17	21	15
E	Water supply;sewerage,waste management		850	931	1133	1125		68728	82406	87394	92390		3680	5275	6649	7598

Viet Nam

ISIC Revision 4		Output (valuation not defined)					Value added at producers' prices				
			(trillions of Vietnamese Dongs)					(trillions of Vietnamese Dongs)			
ISIC	Industry	Note	2010	2011	2012	2013a/	Note	2010	2011	2012	2013
05	Mining of coal and lignite		50.6	63.3	70.2	78.9		20.1	21.7	24.1	24.4
06	Extraction of crude petroleum & natural gas		159.1	183.6	268.4	287.9		99.6	169.3	172.8	151.5
07	Mining of metal ores		5.5	10.5	10.9	11.5		3.0	5.0	5.4	6.1
08	Other mining and quarrying		30.2	32.5	28.4	28.7		11.8	11.6	11.6	12.4
09	Mining support service activities		...	...	...	...		10.3	11.6	17.0	18.2
B	Mining and quarrying		250.5	297.1	384.9	413.8		144.9	219.3	231.0	212.6
35	Electricity, gas, steam & air conditioning		...	...	...	...		153.0	164.8	195.7	222.5
D	Electricity, gas, steam & air conditioning		...	...	...	...		153.0	164.8	195.7	222.5
36	Water collection, treatment and supply		8.9	10.2	11.8	13.8		7.1	6.5	7.6	8.7
37	Sewerage		1.2	1.3	2.4	3.2		1.0	...	1.3	1.5
38	Waste collection,treatment,disposal activities		7.0	7.8	9.9	12.9		3.6	5.8	5.7	7.6
39	Remediation activities		0.3	0.2	0.1	0.2		0.2	0.2	0.1	0.1
E	Water supply;sewerage,waste management		17.5	19.4	24.3	30.2		11.9	14.7	14.7	17.9

a/ Provisional data.

Yemen

ISIC Revision 3		Number of establishments					Number of persons engaged					Wages and salaries paid to employees				
		Note	(number)				Note	(number)				Note	(millions of Yemeni Rials)			
ISIC	Industry		2009	2010	2011	2012		2009	2010	2011	2012		2009	2010	2011	2012
101	Mining and agglomeration of hard coal		...	...	...	...		...	...	...	...		...	...	...	...
102	Mining and agglomeration of lignite		...	...	...	...		...	...	...	...		...	...	...	...
103	Extraction and agglomeration of peat		...	...	...	...		...	...	...	...		...	...	...	...
111	Extraction of crude oil and natural gas		...	...	...	...		...	...	...	...		...	...	...	...
112	Oil, gas service activities		...	...	...	...		...	...	...	...		...	...	...	...
120	Mining of uranium and thorium ores		...	...	...	...		...	...	...	...		...	...	...	...
131	Iron ores		...	...	...	...		...	...	...	...		...	...	...	...
132	Non-ferrous metal ores,excl.uranium,thorium		...	...	...	...		...	...	...	...		...	...	...	...
141	Quarrying of stone, sand and clay		287	...	286	343		3519	...	2899	3462		1311	...	1083	1672
142	Mining and quarrying n.e.c.		6	...	6	...		18	...	24	...		3	...	4	...
C	Total mining and quarrying		293	...	292	...		3537	...	2923	...		1314	...	1086	...
401	Electricity production, collection, distribution		661a/	...	659	721		...	...	...	948		137a/	...	134	235
402	Gas production and distribution		...a/	...	...	...		...	...	...	...		...a/	...	...	...
403	Steam and hot water supply		...a/	...	...	...		...	...	...	...		...a/	...	...	...
410	Water collection, purification, distribution		1797	...	1797	2225		4132	...	4197	5332		638	...	648	2634
E	Total electricity, gas and water supply		2458	...	2456	2946		4692	...	...	6280		775	...	782	2869

a/ 401 includes 402 and 403.

Yemen

ISIC Revision 3		Output (valuation not defined)					Value added (valuation not defined)				
			(millions of Yemeni Rials)					(millions of Yemeni Rials)			
ISIC	Industry	Note	2009	2010	2011	2012	Note	2009	2010	2011	2012
101	Mining and agglomeration of hard coal		...	...	...	...		...	...	...	...
102	Mining and agglomeration of lignite		...	...	...	...		...	...	...	...
103	Extraction and agglomeration of peat		...	...	...	...		...	...	...	...
111	Extraction of crude oil and natural gas		...	...	...	...		...	...	...	...
112	Oil, gas service activities		...	...	...	...		...	...	...	...
120	Mining of uranium and thorium ores		...	...	...	...		...	...	...	...
131	Iron ores		...	...	...	...		...	...	...	...
132	Non-ferrous metal ores,excl.uranium,thorium		...	...	...	...		...	...	...	...
141	Quarrying of stone, sand and clay		16885	...	17232	25624		10777	...	10994	15449
142	Mining and quarrying n.e.c.		21	...	46	...		13	...	33	...
C	Total mining and quarrying		16906	...	17278	...		10790	...	11027	...
401	Electricity production, collection, distribution		3353a/	...	3206	4341		2555a/	...	2436	3073
402	Gas production and distribution		...a/	...	...	...		...a/	...	...	...
403	Steam and hot water supply		...a/	...	...	...		...a/	...	...	...
410	Water collection, purification, distribution		17920	...	18294	23088		11459	...	11638	13011
E	Total electricity, gas and water supply		21274	...	21500	27429		14015	...	14074	16084

a/ 401 includes 402 and 403.